Optical Coherence Tomography and Its Non-medical Applications

Edited by Michael R. Wang

Published in London, United Kingdom

IntechOpen

Supporting open minds since 2005

Optical Coherence Tomography and Its Non-medical Applications
http://dx.doi.org/10.5772/intechopen.81767
Edited by Michael R. Wang

Contributors
Kai Shen, Hui Lu, Sarfaraz Baig, Michael Wang, James Wang, Piotr Targowski, Magdalena Iwanicka, Magdalena Kowalska, Marcin Sylwestrzak, Jun Zhang, Xinyu Li, Shanshan Liang, Hari Nandakumar, Shailesh Srivastava, Luca Fortunato, Nobuhiko Ozaki, Kazumasa Ishida, Tsuyoshi Nishi, Hirotaka Ohsato, Eiichiro Watanabe, Naoki Ikeda, Yoshimasa Sugimoto, Vladimir Zaitsev, Lev Matveev, Alexander Sovetsky, Alexander Matveyev, Chao Zhou, Yongyang Huang, Hao Huang, Zhiyu Jiang, Lanfang Li, Willie Lau, Mohamed El-Aasser, H. Daniel Ou-Yang, Nelida Araceli Russo, Eneas Morel, Jorge Torga, Ricardo Duchowicz, Jianing Yao, Jannick Rolland

First published in London, United Kingdom, 2020 by IntechOpen
IntechOpen is the global imprint of INTECHOPEN LIMITED, registered in England and Wales, registration number: 11086078, 7th floor, 10 Lower Thames Street, London, EC3R 6AF, United Kingdom
Printed in Croatia

British Library Cataloguing-in-Publication Data
A catalogue record for this book is available from the British Library

Additional hard and PDF copies can be obtained from orders@intechopen.com

Optical Coherence Tomography and Its Non-medical Applications
Edited by Michael R. Wang
p. cm.
Print ISBN 978-1-78984-261-6
Online ISBN 978-1-78984-262-3
eBook (PDF) ISBN 978-1-83880-801-3

An electronic version of this book is freely available, thanks to the support of libraries working with Knowledge Unlatched. KU is a collaborative initiative designed to make high quality books Open Access for the public good. More information about the initiative and links to the Open Access version can be found at www.knowledgeunlatched.org

For EU product safety concerns:
IN TECH d.o.o., Prolaz Marije Krucifikse Kozulić 3, 51000 Rijeka, Croatia, info@intechopen.com or visit our website at intechopen.com.

Meet the editor

Michael R. Wang is Professor of the Department of Electrical and Computer Engineering, University of Miami. He received his PhD degree in 1992 from the Department of Electrical Engineering, University of California, Irvine. His research is focused on integrated photonic devices, optical interconnects, holography, lithography, spectral imaging, and optical coherence tomography. He has developed optical coherence tomography systems to support various medical and industrial 3D imaging applications. Dr. Wang has been a principal investigator and/or project leader in many US government-sponsored projects. He has been an invited author and editor on lithography by IntechOpen. He has authored/coauthored more than 200 journal papers, proceedings, and conference presentations. He is a fellow of SPIE, a senior member of OSA, and a member of ARVO.

Contents

Preface

Optical coherence tomography (OCT) is a low-coherence interferometer-based non-invasive non-contact imaging modality that can provide high-resolution subsurface 3D imaging of biological tissues as well as various materials and microstructures. Since its introduction in the early 1990s, OCT has attracted significant attention in research, development, and applications.

There are several OCT system configurations, including time domain OCT, spectral domain OCT, swept source OCT, and full-field OCT. In OCT technology development, there are various reports on scan depth control, axial imaging resolution, lateral imaging resolution, polarization sensitivity and effects, scan imaging techniques, and handheld and other packaging techniques. There are many published papers on its interferometry theory, imaging dynamic range enhancement, and resolution improvements.

The majority of OCT imaging applications have so far been found in medical/biomedical areas. There are several published books related to OCT technologies and medical/biomedical imaging applications. In recent years, OCT has received more and more attention in non-medical applications, including, for example, inspection of polymer films and fibers, fingerprint acquisition, surface and coating evaluations, evaluation of materials and paintings, 3D imaging of microstructures, and evaluation of fastener flushness. In this book we present some recent advancements in OCT technology and non-medical applications.

Michael R. Wang, PhD
Department of Electrical and Computer Engineering,
University of Miami,
Florida, USA

Section 1

Non-Contact 3D Imaging Technology

Chapter 1

Dynamic Range Enhancement in Swept-Source Optical Coherence Tomography

Jun Zhang, Xinyu Li and Shanshan Liang

Abstract

The imaging penetration depth of an optical coherence tomography (OCT) system is limited by the dynamic range of the system. In a common case that signals exceed the dynamic range of a Fourier domain OCT (FDOCT) system, saturation artifacts degrade the image quality. In this chapter, we demonstrate some new cost-effective techniques to improve the dynamic range of a swept-source OCT (SSOCT) system. For example, one method is based on a dual-channel detection technique to enhance the dynamic range by reconstructing the saturated signals due to strong reflection of the sample surface. Another method utilizes a tunable high-pass filter to compensate the attenuation of light signal in deep tissue. It was demonstrated that these techniques can improve the dynamic range of an SSOCT system by more than 10 dB with a low bit-depth analog-to-digital converter.

Keywords: saturation artifacts, dynamic range, dual-channel detection, attenuation compensation, endoscopic OCT, contrast

1. Introduction

As a noninvasive, high-resolution tomographic technique providing cross-sectional and three-dimensional imaging of biological tissue in micrometer scale, optical coherence tomography (OCT) has been widely used in many clinical applications including ophthalmology [1], dermatology, interventional cardiology imaging, airway imaging [2, 3], etc.

Compared to time domain OCT (TDOCT), Fourier domain OCT (FDOCT) can achieve a much higher sensitivity and imaging speed [4]. Using the Fourier domain technique based on a high-speed wavelength swept source, swept-source OCT (SSOCT) is capable of an A-line rate of up to multi-MHz [5] with a simple fiber-based setup, which makes SSOCT attractive in clinical applications especially in endoscopic imaging of internal organs [6, 7].

In spite of all the advantages of SSOCT, there are still some impediments that degrade the image quality especially in in vivo endoscopic studies. One of the challenges is the saturation effects due to the strong signals from highly reflective areas such as surface of internal organs that are commonly lubricated by mucus, catheters and guide wires assembled in endoscopic probes, metallic stent struts and micro-calcifications, etc. Since the incident angle is hard to control in endoscopic OCT imaging, the power of the reflected signal light occasionally exceeds the input range of the detector or analog-to-digital converter (ADC). Fourier transformation of the saturated signal results in a bright line on the tissue surface accompanied with a band of artifacts across tissue depth that degrades the image quality and leads to complete loss of information in the areas with strong artifacts.

One solution is to increase the input range of the detector or ADC especially the latter since the limited bit depth of the ADC is usually the bottleneck of the dynamic range of the system. However, a high-performance ADC with a high bit depth and high sampling frequency is costly. Huang et al. reported a method to correct saturation artifacts by linear interpolation of the signals in adjacent A-lines [8]. However, the interpolation-based reconstruction can only be used in the correction of sparse saturation artifacts. An adaptive optimization technique based on automatic adjustment of the reference power was used to suppress saturation effects in spectral-domain OCT (SDOCT) [9] at the cost of significantly slowing down the frame rate due to complex design and calculations. Wu et al. utilized a multi-exposure spectrum recording method to reduce saturation artifacts in SDOCT [10]. However, the compensation effect was limited by the inaccurate estimation of the multi-exposure signal levels since the ratio of the levels cannot be precisely calibrated. Therefore, a real-time and accurate technique to correct saturation effect suited for SSOCT systems especially in endoscopic imaging is still absent.

An alternative design based on a dual-channel detection technique was presented to suppress the saturation artifacts [11]. The detected signal was split into the two channels with the ratio of the signal levels precisely calibrated. The high-level signal was used to reconstruct OCT images, and the low-level signal was used to correct the saturated signal in the case that the high-level signal exceeds the input range of the system. This technique allows for a simple and cost-effective suppression of saturation artifacts in endoscopic SSOCT without the need of decreasing the incident power.

Another impediment that degrades the image quality of OCT is that image contrast decays drastically with imaging depth due to strong attenuation of light in biological tissues [12]. Chang et al. reported a method to compensate OCT signal attenuation in depth by adaptively deriving a compensation function for each A-scan line [13]. Hojjatoleslami et al. proposed an enhancement algorithm for attenuation compensation to improve the image quality in the structures at deeper levels [14]. Zhang et al. built a dual-band FDOCT system and developed an algorithm to compensate depth-related discrepancy and attenuation [15]. An alternative approach of compensating attenuation by performing extraction of optical scattering parameters was presented by Anderson et al. [16, 17]. Girard et al. developed a series of algorithms that can be applied to compensate light attenuation and enhance contrast in both time and spectral-domain OCT images [18]. However, these algorithm-based approaches require a prohibitive number of computations and are not practical for real-time imaging. Recently, Li et al. combined a tunable high-pass filter with a dual-channel ADC to compensate signal decay in deep tissue in real time [19]. Since signal frequency represents the depth in SSOCT imaging, low-frequency signal in one channel that is filtered out by a high-pass filter and then combined with the signal in the other channel can be used to reconstruct a high-contrast image in both surface and deep area of the tissue.

2. Correction of saturation effects

The schematic diagram of the SSOCT system for correction of saturation effects is shown in **Figure 1** [11]. A swept source at 1310 nm with a bandwidth of 87 nm, a sweep frequency of 100 kHz, and an output power of 20 mW were used as the light source. The input light was split by a 1 × 2 coupler into the sample and reference arms, respectively. In the sample arm, a 1.3 mm proximal scanning endoscopic probe was employed for three-dimensional imaging. The helical scanning probe was driven by a rotary motor with a rotational rate of 50 rounds/seconds and a stepper motor translational stage with a pulling-back speed of 1 mm/second, respectively. By using a phase-resolved algorithm to computationally compensate the dispersion generated by the endoscope optics, the SSOCT system is capable of an axial resolution of 8 μm in the tissue and a lateral resolution of 20 μm, respectively. The total reference power was set to be 25 μW for optimization of the system sensitivity. In the detection arm, a balanced detector with the noise level comparable to the quantization noise of the ADC was used. In order to compensate saturation artifacts, the interference signal was divided into two paths by a broadband power divider and then digitized by a 12-bit two-channel ADC. In each channel, 1024 samples were acquired using the k-clock from the laser source as an external clock signal. The splitting ratio of the power divider was accurately calibrated by utilizing a high-performance oscilloscope. The signal collected by the high-level channel (ChA) was used for OCT imaging. To detect saturation in ChA, a threshold of the low-level signal in ChB was set to be equal to the input range divided by the splitting ratio of the power divider. Hence, the saturated signal in ChA over the maximum input range can be reconstructed with the signal spontaneously detected in ChB. By multiplying the splitting ratio with the signal in ChB, the saturated signal due to strong reflection was compensated as shown in **Figure 2A**. The corresponding

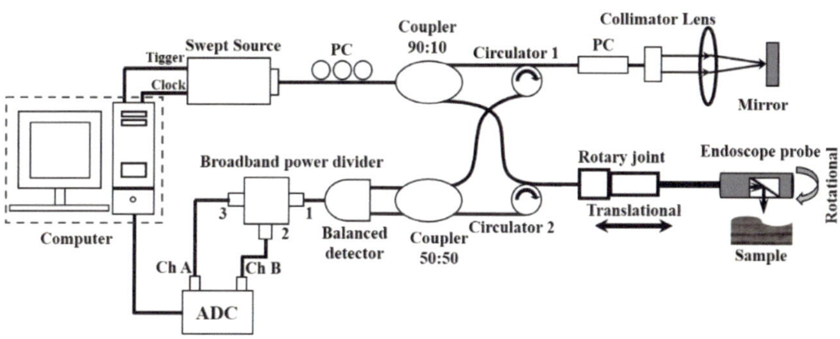

Figure 1.
Schematic of the SSOCT system for saturation correction: PC, polarization controller; ADC, analog to digital converter [11].

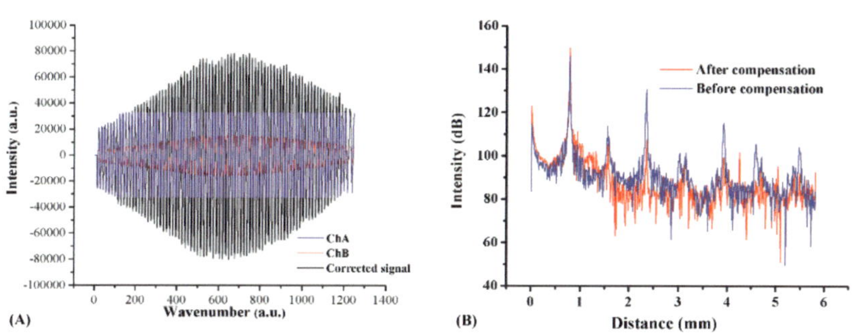

Figure 2.
(A) Interference signals recorded with ChA (blue) and ChB (red). Black line denotes corrected signals in ChA after compensation with signals in ChB. (B) Fourier transforms of signals before and after compensation.

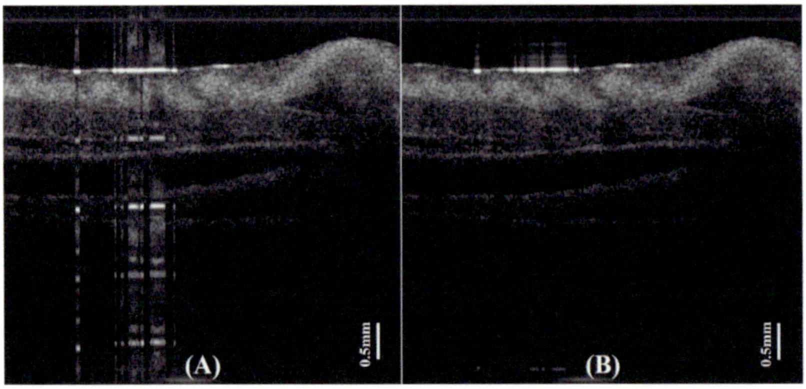

Figure 3.
OCT images of porcine airway before (A) and after (B) correction [11].

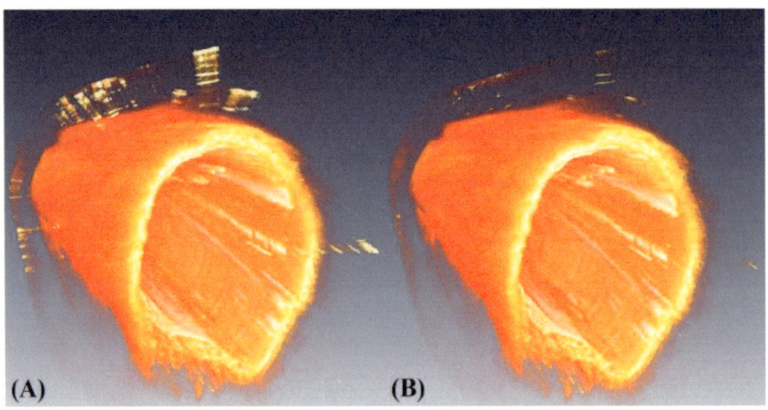

Figure 4.
3D endoscopic OCT images of porcine airway before (A) and after (B) correction.

artifact peaks in depth domain after Fourier transformation were significantly suppressed (**Figure 2B**).

To evaluate the system's capacities of imaging tissues with high reflectivity, a section of porcine upper airway tissues was imaged using the saturation-correction system. OCT imaging was processed on a graphical processing unit (GPU) featuring a multithreaded real-time data acquisition, image processing, and display at the rate of 50 frames/second with 2000 A-lines in each frame. As illustrated in **Figure 3**, the structures hidden inside the bright vertical lines were revived through significant suppression of saturation artifacts [11].

Construction of 3D data sets from 500 B scan utilized a commercial software package. As shown in **Figure 4**, the artifacts were removed with this technique resulting in a clean 3D reconstruction of endoscopic OCT imaging.

3. Compensation of signal attenuation

Figure 5 illustrates the averaged intensity of 1000 A-lines in SSOCT imaging of the human skin showing that signal frequency represents depth in the sample. Hence, attenuated signals in high frequency can be compensated in frequency domain by using a tunable high-pass filter to filter out the low-frequency signal. **Figure 6** shows the frequency response of the high-pass filter with a cutoff frequency set to be 13 MHz.

Figure 7 shows the SSOCT system for attenuation compensation [19]. A swept source with a center wavelength of 1310 nm, a bandwidth of 108 nm, a sweep frequency of 50 kHz, and an output power of 20 mW were used as the light source. The light was split into the sample arm and the reference arm by a 90:10 coupler. The light back-scattered/back-reflected from the reference mirror and sample arm was redirected by two circulators and detected by a balanced detector. The detected signal was divided into two channels of an ADC by a directional coupler with the ratio of 1:5. A high-pass filter was utilized to remove the low-frequency component from the higher-level signal in ChB so that the signal intensity in ChB is close to that in ChA. Since the higher-frequency signals experience stronger attenuation, signals in ChB and ChA can be used to reconstruct structure in deep tissue and surface, respectively. The signals in two channels were combined in real time after being digitized by the ADC.

The human finger was imaged to test the capability of the system to improve the image contrast. As illustrated in **Figure 8**, OCT image after compensation shows an obvious enhancement of contrast in deep area.

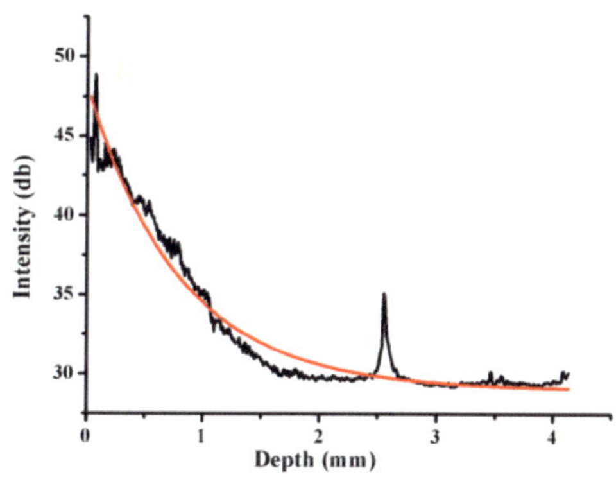

Figure 5.
Averaged intensity of 1000 A-lines in human skin imaging.

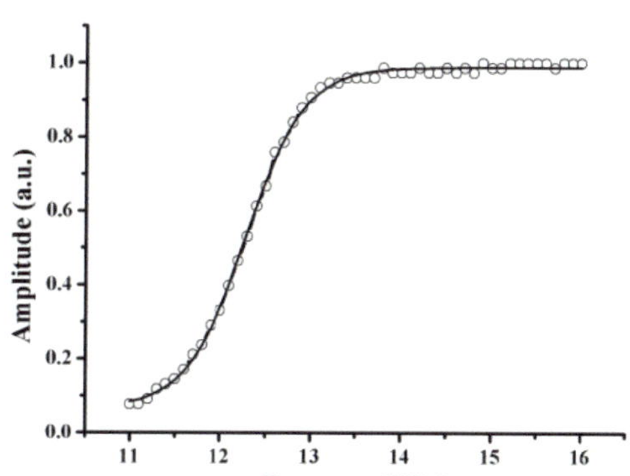

Figure 6.
Frequency response of the high-pass filter.

Porcine upper airway imaging shown in **Figure 9** demonstrated that this method can be used to improve OCT image quality effectively [19].

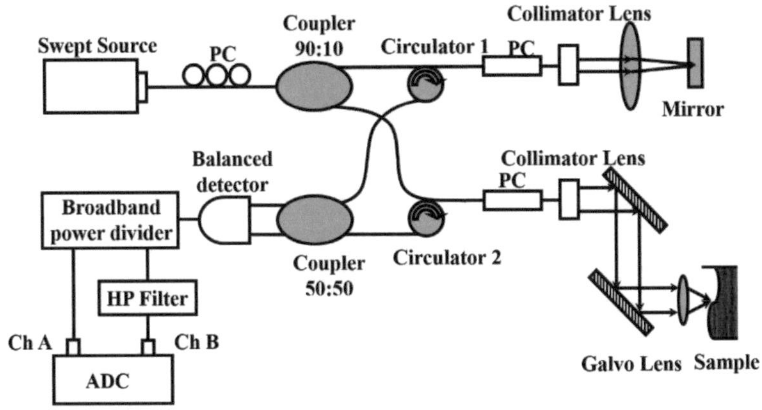

Figure 7.
Schematic of the SSOCT system for attenuation compensation: ADC, analog-to-digital converter.

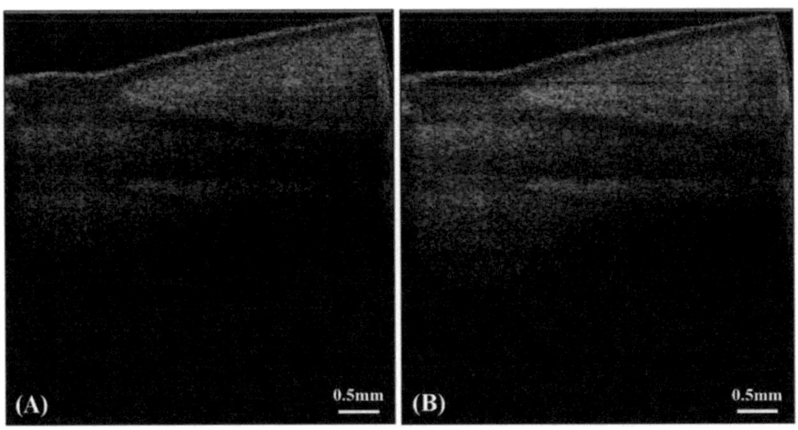

Figure 8.
OCT images of the human finger before (A) and after (B) compensation.

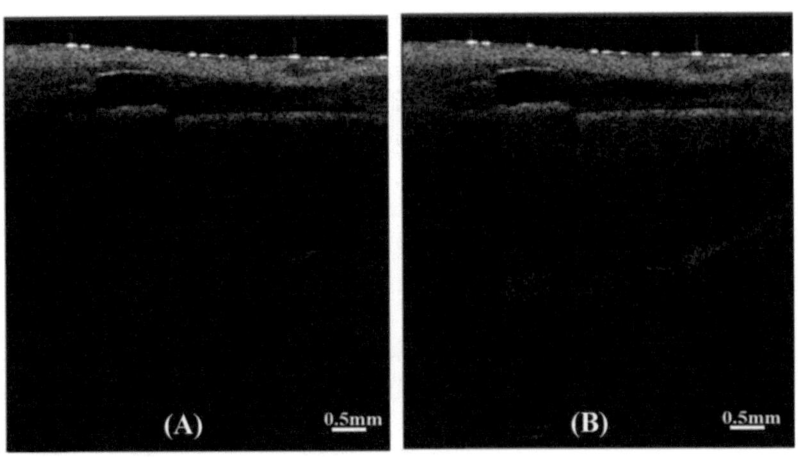

Figure 9.
OCT images of porcine upper airway before (A) and after (B) compensation [19].

4. Conclusions

In clinical applications such as dermatological imaging, the reflectivity of dry tissue is less than 10^{-4}. However, in endoscopic imaging applications, the reflectivities of mucous fluid, catheters, and stent struts usually exceed 10^{-2} or more, resulting in saturation artifacts in OCT images. Increasing the number of bits of an ADC could reduce saturation effects, however requiring complex and costly design. The dual-channel-based saturation-correction approach provides a simple and cost-effective method to solve this problem. The results showed this technique effectively suppresses saturation artifacts especially in endoscopic OCT imaging.

Due to strong attenuation of light in biological tissues, OCT signal decreased dramatically with the penetration depth. The attenuation compensation approach combining a tunable high-pass filter with a dual-channel ADC enhances the contrast of OCT images in a deeper region effectively. Human finger and porcine upper airway imaging demonstrated that high-quality image can be obtained with this method.

In conclusion, the dynamic range of an SSOCT system can be improved by more than 10 dB with a low bit-depth analog-to-digital converter by using these techniques.

Author details

Jun Zhang*, Xinyu Li and Shanshan Liang
School of Electronics and Information Technology, Sun Yat-sen University, Guangzhou, China

*Address all correspondence to: junzhang8819@gmail.com

References

[1] Costa RA, Skaf M, Melo LAS, et al. Retinal assessment using optical coherence tomography. Progress in Retinal and Eye Research. 2006;**25**(3):325-353. DOI: 10.1016/j.preteyeres.2006.03.001

[2] Sinclair H, Bourantas C, Bagnall A, et al. OCT for the identification of vulnerable plaque in acute coronary syndrome. JACC: Cardiovascular Imaging. 2015;**8**(2):198-209. DOI: 10.1016/j.jcmg.2014.12.005

[3] Milne S, King GG. Advanced imaging in COPD: Insights into pulmonary pathophysiology. Journal of Thoracic Disease. 2014;**6**(11):1570-1585. DOI: 10.3978/j.issn.2072-1439.2014.11.30

[4] Leitgeb R, Hitzenberger CK, Fercher AF. Performance of Fourier domain vs. time domain optical coherence tomography. Optics Express. 2003;**11**(8):889-894. DOI: 10.1364/OE.11.000889

[5] Huo T, Wang C, Zhang X, et al. Ultrahigh-speed optical coherence tomography utilizing all-optical 40 MHz swept-source. Journal of Biomedical Optics. 2015;**20**(3):30503. DOI: 10.1117/1.JBO.20.3.030503

[6] Gabriel S. In vivo endoscopic optical coherence tomography of the human gastrointestinal tract - toward optical biopsy. Endoscopy. 2000;**32**(10):743-749. DOI: 10.1055/s-2000-7711

[7] Lee SW, Heidary AE, Yoon D, et al. Quantification of airway thickness changes in smoke-inhalation injury using in-vivo 3-D endoscopic frequency-domain optical coherence tomography. Biomedical Optics Express. 2011;**2**(2):243-254. DOI: 10.1364/BOE.2.000243

[8] Huang Y, Kang JU. Real-time reference A-line subtraction and saturation artifact removal using graphics processing unit for high-frame-rate Fourier-domain optical coherence tomography video imaging. Optical Engineering. 2012;**51**(7):3203. DOI: 10.1117/1.OE.51.7.073203

[9] Kim JH, Han JH, Jeong J. Adaptive optimization of reference intensity for optical coherence imaging using galvanometric mirror tilting method. Optics Communications. 2015;**351**:57-62. DOI: 10.1016/j.optcom.2015.04.041

[10] Wu CT, Tsai MT, Lee CK. Two-level optical coherence tomography scheme for suppressing spectral saturation artifacts. Sensors. 2014;**14**(8):13548-13555. DOI: 10.3390/s140813548

[11] Li X, Liang S, Zhang J. Correction of saturation effects in endoscopic swept-source optical coherence tomography based on dual-channel detection. Journal of Biomedical Optics. 2018;**23**(3):1. DOI: 10.1117/1.JBO.23.3.030502

[12] Jansonius NM, Cervantes J, Reddikumar M, et al. Influence of coherence length, signal-to-noise ratio, log transform, and low-pass filtering on layer thickness assessment with OCT in the retina. Biomedical Optics Express. 2017;**7**(11):4490. DOI: 10.1364/BOE.7.004490

[13] Chang S, Flueraru C, Mao Y, et al. Attenuation compensation for optical coherence tomography imaging. Optics Communications. 2009;**282**(23):4503-4507. DOI: 10.1016/j.optcom.2009.08.030

[14] Hojjatoleslami A, Avanaki MRN. OCT skin image enhancement through attenuation compensation. Applied Optics. 2012;**51**(21):4927-4935. DOI: 10.1364/AO.51.004927

[15] Zhang M, Ma L, Yu P. Dual-band Fourier domain optical coherence tomography with depth-related compensations. Biomedical Optics Express. 2014;**5**(1):167. DOI: 10.1364/BOE.5.000167

[16] Thrane L, Frosz MH, Jørgensen TM, et al. Extraction of optical scattering parameters and attenuation compensation in optical coherence tomography images of multilayered tissue structures. Optics Letters. 2004;**29**(14):1641-1643. DOI: 10.1364/OL.29.001641

[17] Thrane L, Yura HT, Andersen PE. Analysis of optical coherence tomography systems based on the extended Huygens-Fresnel principle. Journal of the Optical Society of America A. 2000;**17**(3):484-490. DOI: 10.1364/JOSAA.17.000484

[18] Girard MJA, Strouthidis NG, Ethier CR, et al. Shadow removal and contrast enhancement in optical coherence tomography images of the human optic nerve head. Investigative Ophthalmology & Visual Science. 2011;**52**(10):7738. DOI: 10.1167/iovs.10-6925

[19] Li X, Liang S, Zhang J. Contrast improvement for swept-source optical coherence tomography image of sub-surface tissue. In: Optical Coherence Tomography and Coherence Domain Optical Methods in Biomedicine XXI. International Society for Optics and Photonics. 2017;**10053**:100532R. DOI: 10.1117/12.2251650

Chapter 2

Multi-Frame Superresolution Optical Coherence Tomography for High Lateral Resolution 3D Imaging

Kai Shen, Hui Lu, Sarfaraz Baig and Michael R. Wang

Abstract

We report that high lateral resolution and high image quality optical coherence tomography (OCT) imaging can be achieved by the multi-frame superresolution technique. With serial sets of slightly lateral shifted low resolution C-scans, our multi-frame superresolution processing of these special sets at each depth layer can reconstruct a higher resolution and quality lateral image. Layer by layer repeat processing yields an overall high lateral resolution and quality 3D image. In theory, the superresolution with a subsequent deconvolution processing could break the diffraction limit as well as suppress the background noise. In experiment, about three times lateral resolution improvement has been verified from 24.8 to 7.81 μm and from 7.81 to 2.19 μm with the sample arm optics of 0.015 and 0.05 numerical apertures, respectively, as well as the image quality doubling in dB unit. The improved lateral resolution for 3D imaging of microstructures has been observed. We also demonstrated that the improved lateral resolution and image quality could further help various machine vision algorithms sensitive to resolution and noise. In combination with our previous work, an ultra-wide field-of-view and high resolution OCT has been implemented for static non-medical applications. For *in vivo* 3D OCT imaging, high quality 3D subsurface live fingerprint images have been obtained within a short scan time, showing beautiful and clear distribution of eccrine sweat glands and internal fingerprint layer, overcoming traditional 2D fingerprint reader and benefiting important biometric security applications.

Keywords: optical coherence tomography, lateral resolution, superresolution, 3D imaging, microstructure, fingerprint identification

1. Introduction

Optical coherence tomography (OCT) [1, 2] is an advanced non-contact 3D imaging technique, providing subsurface cross-sectional tomographic images. It offers deeper penetration depth [3] and larger scan area [4] than confocal microscope imaging [5] as well as higher resolution [3] than ultrasonic imaging [6]. It is thus widely utilized in 3D imaging of eyes [7, 8], skins [9–12], blood vessels [13], cartilages [14], and numerous biomedical applications.

With non-contact and non-invasive advantages, OCT has significant medical applications. There is also a huge potential of OCT for many non-biomedical applications that demands non-destructive testing and evaluations in micron scale resolutions [15]. For example, there is no preparatory steps for OCT sample imaging, instead of gold coating for SEM imaging; no coupling media as required for ultrasound imaging; no special safety precautions like X-ray. Also, the near infrared light source in OCT usually has no photo reactions with most materials, very safe for quality testing of damage in silica [16], glass-fiber reinforced polymer samples [17], strained polymer samples [18], microstructures [19–23], papers [24], oil paintings [25], film coatings [26], fastener flushness [27], and so on. Besides, the successful detection of embedded and hidden structures is another potential of OCT for security applications, such as 3D fingerprint identification defending against spoofing attack with fake fingerprints [28–31]. However, compared with other imaging techniques such as microscopy and confocal microscopy, the low lateral resolution and high speckle noise restrict the OCT becoming a competitive imaging tool in some non-biomedical areas highly relying on *en-face* lateral image quality.

OCT imaging has two distinct resolutions namely axial resolution in the depth direction and lateral resolution in the *en-face* plane like microscopy. The axial resolution regards to the coherence length of the light source and thus can be improved by supercontinuum light [32] or extended broadband superluminescent diode (SLD) [33]. The lateral resolution is mainly restricted by diffraction limit [34], lateral sampling rate [35] and background noise [36]. The diffraction limit is the minimum focused spot size, determined by the numerical aperture (*NA*) of the OCT sample arm optics. Although a high *NA* optics could achieve a smaller focused beam spot size on the sample, the quick divergence of the beam size out of the focal plane reduces the *depth of focus* (*DOF*) of the OCT system, losing its main advantage over confocal microscope. Higher *NA* also limits the lateral field-of-view (FOV) due to the rapid off-axis degradation of the focusing performance, explained in our previous work [37, 38]. Therefore, it is crucial to overcome the complex trade-off among lateral resolution, axial *DOF*, and lateral FOV in the OCT imaging.

Adaptive optics (AO), an astronomical telescope technique, has been adopted in OCT systems to correct aberration wave front and thus improves the lateral resolution [39]. Except the high cost and a very limited FOV (maximum 1×1 mm^2) [40, 41], AO technique in principle is to recover the original lateral resolution of OCT, which however is blurred by human eyes. Thus, it is not suitable for non-ophthalmic imaging like skins due to scattering blurring. A virtually structured detection (VSD) method [42] was reported to improve the lateral resolution by adding an electro-optic phase modulator (EOPM) in the reference arm. The EOPM shifts the light phase with multiple of $\pi/2$, and then the VSD algorithm fuses four phase shifted A-scans to one, achieving resolution doubling. It is a time consuming (taking ~40 s for each image frame) technique which is infeasible for *in vivo* imaging and 3D imaging. Robinson et al. [43] register four sparse scanned summed voxel projection (SVP) of retina images to reconstruct a higher density *en-face* image in *y*-axis to improve the resolution and reduce motion errors, while the quality improvement does not overcome the traditional high density scan images. Digital image deconvolution processing is a potential technique to break diffraction limit and improve the resolution [44–47]. The estimation of the ground true lateral point spread function (PSF) of the system is however very difficult and the actual PSF may be different in different samples and at different depth layers.

Background noise is another factor degrading the resolution and image quality of OCT systems. Different from white noise, the structure related speckle noise in OCT imaging is difficult to be suppressed by the multi-frame averaging [48, 49]. Szkulmowski et al. [36] introduced an interesting averaging algorithm with multiple shifted B-scans to remove the speckle noise. However, this approach introduces new ghost patterns in *in vivo* imaging, such as multiple ghost fingertip patterns in the output image, due to averaging multiple B-scans in different positions. Besides, simple averaging shifted images may penalize the high frequency signals and degrade the resolution. And the longer B-scan time is impractical for 3D imaging.

Lateral sampling rate [35] in scan-based OCT imaging is termed the scan matrix density. According to Nyquist-Shannon sampling theorem, the sampling frequency should double the sample frequency at least. Although increasing the scan matrix density could improve the lateral resolution, this method is at the expense of longer scan time and not suitable for time sensitive applications, such as *in vivo* imaging of fingerprint. Besides, high scan density cannot overcome the diffraction limit and reduce background noises.

In this book chapter, we report an effective multi-frame superresolution technique to significantly improve the lateral resolution and image quality of OCT without adoption of extra hardware and higher *NA* optics. Through adjustment of galvanometer scanners to introduce slightly shifts among different sparse sampled C-scans, the superresolution processing is then applied to generate a three times higher lateral resolution image with suppressed background noise, demonstrated by imaging a standard resolution target. The remarkable improvement of 3D *in vitro* imaging has been observed in a microstructure sample with 2–3 μm scale features. The image stitching technique helps us to reconstruct an ultrawide FOV and high lateral resolution 3D image. For *in vivo* imaging, the image registration method is used to estimate the unknown random shifts among different C-scans. The subsequent superresolution processing demonstrates high quality 3D and subsurface *in vivo* images of fingerprint, benefiting various security applications.

2. OCT system and superresolution principle

2.1 SD-OCT and lateral resolution limit

Our spectral domain optical coherence tomography (SD-OCT) (BIOptoscan OS-186, New Span Opto-Technology) is one kind of the most popular OCT systems in ophthalmic clinic applications, as schematically shown in **Figure 1**. It sends a broadband light from the SLD to a 2 × 2 optical fiber coupler. The SLD has a center wavelength λ_o of 860 nm and spectral bandwidth $\delta\lambda$ of 100 nm (IPSDW0822–0314, InPhenix). One split beam is sent to the reference arm that is focused to a mirror and then reflected back to the fiber coupler. The other split beam in the sample arm is focused to the measurement sample and laterally scanned by a pair of galvanometer scanner mirror. The scattering signals from different depth layers of the sample collected by lens are sent back to the fiber coupler to interfere with the return beam from the reference arm, generating spectral interference patterns that are imaged by the optical spectrometer for computer signal processing. Each scattering depth would result in a near sinusoidal interference pattern in the frequency domain. The final spectral image looks complex due to mixing of all interference patterns with different periods from all sample depth layers. A fast Fourier transform processing of the mixed interference pattern in the frequency domain can beautifully retrieve a series interface layers inside the sample within the depth range of the SD-OCT, set by the combination of sample arm optics *DOF* and the spectral resolution of the optical spectrometer. This above processing yields the A-scan, the depth intensity profile $I(z)$ of one point in the lateral plane. Through galvanometer scanning in the transverse x axis, we obtain the B-scan x-z intensity image $I(x,z)$. By galvanometer

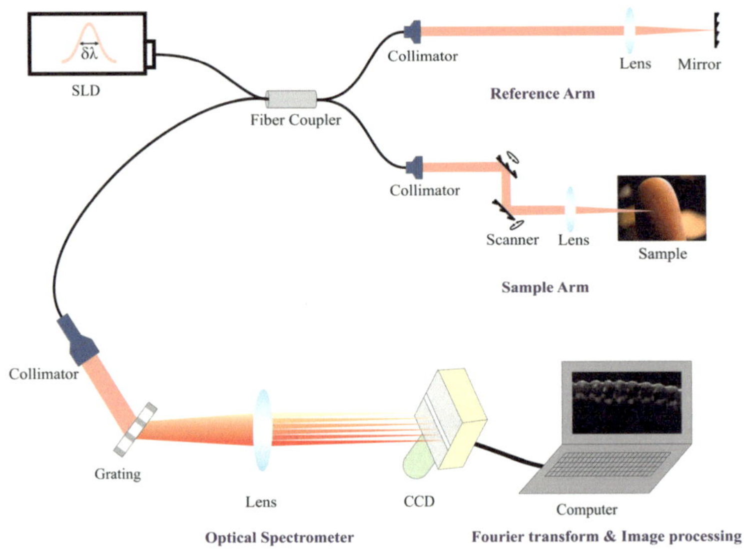

Figure 1.
Schematic configuration of our SD-OCT system.

scanning in both the transverse x and y axes line by line, we have the C-scan x-y-z intensity 3D image $I(x, y, z)$.

As discussed earlier, the focused beam spot size at full width half maximum (FWHM) of the SD-OCT imaging system is mainly limited by the NA of the sample arm optics [34] as

$$\delta x = 0.37\frac{\lambda_0}{NA}. \tag{1}$$

Here, NA is the ratio of the input collimated beam radius to the focal length of the sample arm lens in the air. The axial DOF is determined by [34].

$$DOF_{axial} = \frac{0.565 \cdot \lambda_0}{\sin^2\left[\frac{\sin^{-1}(NA)}{2}\right]}. \tag{2}$$

Given our collimated beam diameter of ~3 mm, the corresponding focusing spot size and axial depth range are listed in **Table 1** for a few common lenses' focal lengths. Here, before reaching spectrometer limitation, the axial depth range is set by the axial DOF.

Focal length (mm)	Beam spot size (μm)	Axial depth range (mm)	Lateral FOV (μm)
10	2.12	0.086	
19	4.03	0.31	500 × 500
30	6.36	0.78	1400 × 1400
100	21.21	8.64*(2.86)	12,000 × 12,000

**Our present optical spectrometer for the SD-OCT supports a maximum depth range of 2.86 mm in the air.*

Table 1.
Four different lenses with their calculated beam sizes, as well as axial depth range and lateral FOV.

Also, the NA of optical system will influence the effective lateral FOV. In theory, the lateral FOV of the OCT is simply given [50] as:

$$FOV_{lateral} = 2 \cdot f \cdot \tan \theta_{max}. \tag{3}$$

Here, f is the focal length of the sample arm lens and θ_{max} depends on both the radius of the lens and the acceptable maximum off-axis scanning angle of the galvanometer scanners.

Typically, the acceptable maximum galvanometer scanning angle θ_{max} could be quite large. However, due to off-axis focused beam aberration and Petzval field curvature [38], the FOV is quite limited. The displacement Δx of an image point at height h_i on the Petzval surface from the paraxial image plane is given by [51]:

$$\Delta x = \frac{h_i^2}{2} \sum_{j=1}^{m} \frac{1}{n_j f_j} \tag{4}$$

Here, n_j and f_j are the indices and focal lengths of the m thin lenses forming the system. This equation implies that the Petzval surface is an unaltered value by changes in position or shapes of lenses and stops, but inversely proportional to the focal lengths. Detailed illustration and explanation could refer to our previous work [38]. Usually the field curvature from high NA lenses would rapidly blur the off-axis image and degrade the image quality at the edge. We thus need to find a suitable FOV with an acceptable off-axis image quality reduction for the selected sample lenses.

In order to quantify the influence of the field curvature to the focusing performance, we simulate the off-axis focusing degradation of three common lenses (Thorlabs, AC254-030-B, AC127-019-B, and AC254-100-B) by ZEMAX software, with two of them shown in **Figure 2**. We see that the focal spot size of 30 mm focal length lens remains almost unchanged when off-axis distance is less than 300 μm. With off-axis distance larger than 800 μm, the focusing degradation becomes obvious. At off-axis distance of 1000 μm, the diameter of focal spot size (measured at FWHM of the peak) is ~16% larger than that at the center position. The increased off-axis focused beam spot size would significantly degrade the OCT lateral resolution. With 1400 × 1400 μm^2 areal scan imaging (700 μm off-axis distance), the lateral resolution at the image edge is considered acceptable.

For 19 mm focal length lens (Thorlabs, AC127-019-B) simulated in **Figure 2**(B), a 500 μm off-axis distance would lead to ~39% larger in focal spot size. Thus, the optimized single C-scan FOV has to be limited to 500 × 500 μm^2 area (250 μm off-axis distance), only one eighth of the 30 mm focal length lens. For 100 mm focal length lens (AC254–100-B), the image quality is usually acceptable in a 1.2 × 1.2 cm^2 [38] (not shown here), but losing lateral resolution due to large spot size as we discussed above. In practice, unavoidable spherical and coma aberrations would further degrade the image quality.

Clearly, smaller focused beam spot size would improve the lateral resolution but at the expense of the DOF and lateral FOV. For example, although a lens with 10 mm focal length would provide the smallest focused spot size of 2.12 μm, its poorest axial DOF of 0.086 mm makes the SD-OCT incompetent to the confocal microscopy which maximally provides a depth range of about 0.2 mm [3]. Also, the ultrashort focal length will restrict the effective lateral FOV to 200 × 200 μm^2 due to rapid off-axis degradation. With a 100 mm focal length lens, the focused beam spot size is about 21 μm, verified by using a laser beam profiler (BP-5.0, New Span Opto-Technology). Although this focused beam can offer the long depth range of 2.86 mm set by the spectrometer as well as 1.2 × 1.2 cm^2 lateral FOV, its large spot

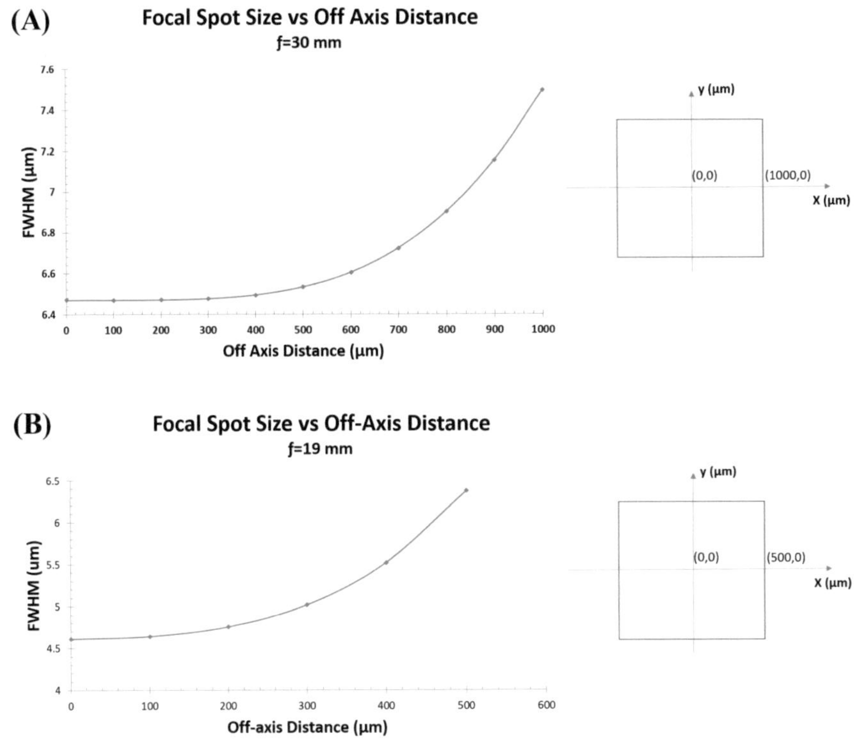

Figure 2.
The curve shows the focus spot size change with the off-axis distances (radius). The scan coordinate is shown at right. Thorlabs AC254-030-B 30 mm (A) and AC127-019-B 19 mm (B) focal length lenses are used in these simulations.

size does not provide desirable lateral resolution, not suitable for imaging of fine structures. Therefore, without special and high cost hardware design and improvement to overcome diffraction limitation, we should consider image processing method to improve the lateral resolution of SD-OCT. If the lateral resolution can be improved to several µm with great image quality and maintain the predominant depth range of 2.86 mm, the processing based lateral resolution improvement technique could benefit security imaging applications such as sub-surface fingerprint reader. Using a 30 mm focal length lens, our goal is to improve the lateral resolution to ~2 µm to approach that of confocal microscope which could benefit micron scale structural imaging.

2.2 Improving lateral resolution by high density scanning

For a large FOV image with short scan time, the focused beam spot scan matrix is usually set to one by one without spot positional overlapping, as illustrated in **Figure 3(b)**, like spatially separated pixel array in an image sensor. The presence of spot spacing results in under sampling and loss of spatial image features. Without demanding smaller focused beam spot size for preserving a long *DOF* and a large FOV, a double or higher density scan matrix with partial scan beam spot overlapping could improve the SD-OCT lateral resolution to some extends explained in Nyquist-Shannon sampling theorem, but at the expense of reducing the lateral FOV, as illustrated in **Figure 3(c)**. Besides the FOV reduction, this high-density scanning method has its resolution limitation and cannot suppress the background noise, discussed in Section 3. In this chapter, the low density scanning

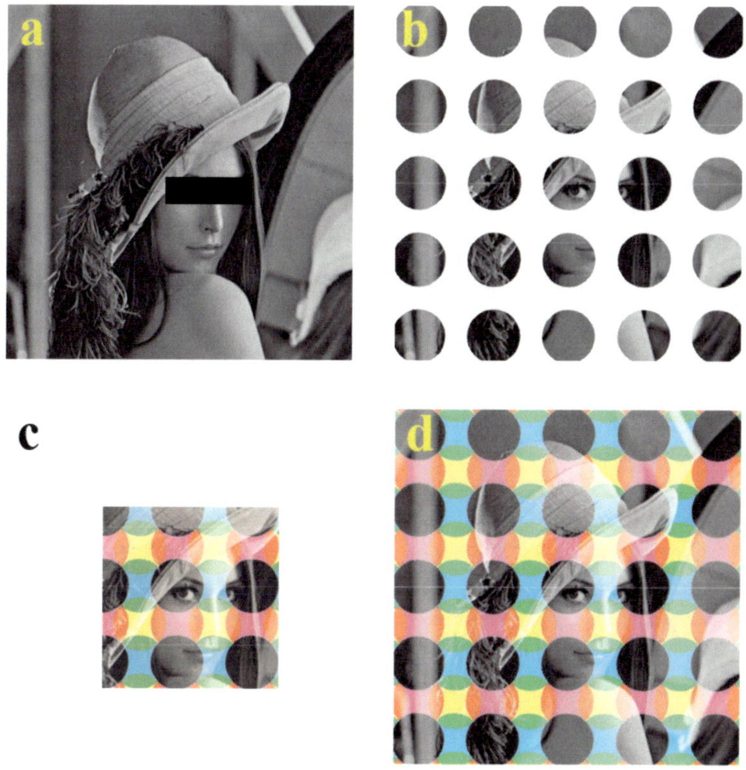

Figure 3.
An illustration image of a high resolution target (a) and its output result under a low density scan imaging (b). (c) 4-time higher density scan with the same scan matrix size as (b) but half spot overlapping of adjacent ones yields higher resolution image while with a reduced FOV. (d) 4-time larger scan matrix leads the same scan density as (c) and the same FOV as (b), but 4-time longer scan time. Here, we apply four colors to show overlapping of adjacent scan spots.

means one by one scan array without nearby spot overlapping and the high density scanning means adjacent scan spots would be partially overlapped. For example, four times high density scanning means each scan spot have half spot overlapping with four neighboring ones (top, bottom, left and right).

In order to avoid FOV reduction, a larger C-scan matrix may be applied to the sample sacrificing the scan time as shown in **Figure 3(d)**, which however is unacceptable for time sensitive *in vivo* imaging of live tissues due to random tissue motion and vibration during the long scan time. To illustrate the temporal motion of live tissue, we performed 100 repeated sparse B-scans of 128-spot to image a human skin. The 100 sets of such B-scan without scan spot overlapping completed within 0.55 s and the fast Fourier transform was calculated later. The comparison of the 1st and the 100th B-scan images shows no observable image shifts as shown in **Figure 4(a)**, demonstrating that a 100×128 or 128×128 (we also verified) C-scan is fast enough for typical live tissue *in vivo* SD-OCT imaging without concerning motion errors in one C-scan. Similarly, we performed 100-frame repeated 512-spot B-scans with the same FOV as above and compared the 1st and the 100th images as shown in **Figure 4(b)**. We observed obvious image positional shifts during the 100×512 scan period indicating some tissue motion during this period. These two experiments were repeated several times with similar results. The scanner optics was held steady during the image acquisition. This indicates that 100×512 C-scan, not so high density, is already inadequate for reliable *in vivo* imaging of live skin tissues owing to live body motion and vibration during the long scan time. Needless

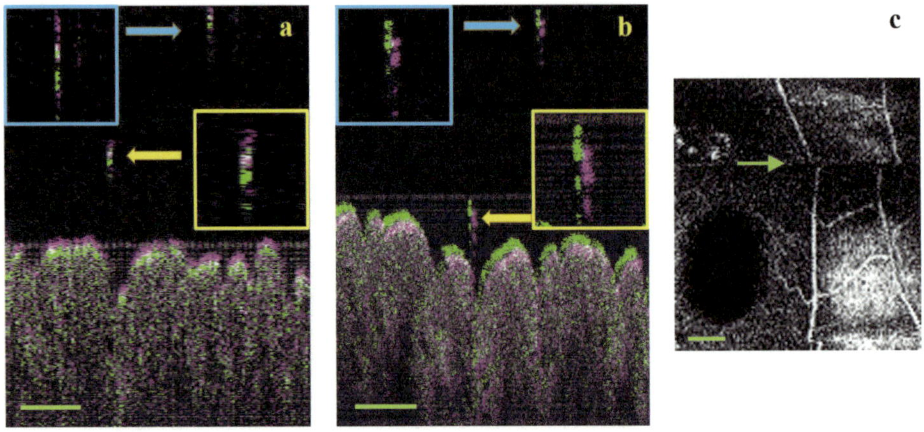

Figure 4.
SD-OCT in vivo *B-scan on a live human skin with 100 repeated frames of (a) 128-spot density and (b) 512-spot density. Same FOV is applied to the two scans. We then overlapped the 1st and the 100th frames to compare the differences by green and magenta colors. (a) Due to the short scan time, the horizontal positions of the hair (in the blue and yellow squares) between the 1st and the 100th images remain almost unchanged. (b) There are obvious shifts between the first and the last images due to slower 512-spot B-scan, in which tissue motion occurs during the scan. (c) The green arrow indicates the discontinued line of the two micro-vessels caused by eye motion during the C-scan. 100 mm focal length lens was used in these experiments. The scale bars in (a), (b), and (c) are 500 μm.*

to say, *in vivo* tissue image misalignments are expected in 512 × 512 or 1024 × 1024 C-scans due to much longer scan time. Take the retina C-scan imaging as an extreme example, unintended eye quick motions can result in clear image artifacts and misalignments, indicated at the green arrow line in **Figure 4(c)**. Thus, the most reliable way for *in vivo* imaging is to scan the sample as fast as possible, avoiding any motion artifacts and errors in one C-scan set. In our experiment, a 128 × 128 C-scan within 0.7 s acquisition time could effectively prevent most motion errors, guaranteeing the data reliability in one C-scan. For unavoidable motion shifts among multiple *in vivo* C-scan sets, an image registration method will be used to align them.

2.3 Improving lateral resolution by the multi-frame superresolution for *in vitro* imaging

The multi-frame superresolution is an image processing technique, studying image degradation models (such as optical blur, motion effect, down-sampling, and additive noise) and then recovering the high resolution image from multiple low quality images based on the superresolution algorithm and the sub-pixel information differences among these images, overcoming the resolution limit of the hardware. **Figure 5** illustrates how these effects result in a low quality image during the conventional camera image acquisition. To recover the high resolution image, a series reversed methods such as up-sampling, motion/pixel shift estimation and compensation, deconvolution, and denoising are applied.

In a SD-OCT system, the main degradations of an ideal lateral image $S(x, y)$ is the optical blurring $H(x, y)$ caused by the lateral PSF of OCT optics and the down-sampling $\downarrow$ due to the sparse scan matrix and the large focused beam spot size. Each isolated focused beam spot is treated as a pixel in conventional camera imaging. The motion effect can be generally ignored when imaging non-biomedical samples such as microstructures [19, 37] since both the sample arm and the samples are stable. The motion effect should be considered for imaging of *in vivo* tissue such as fingerprint [31, 38] due to potential live body vibrations. According to the

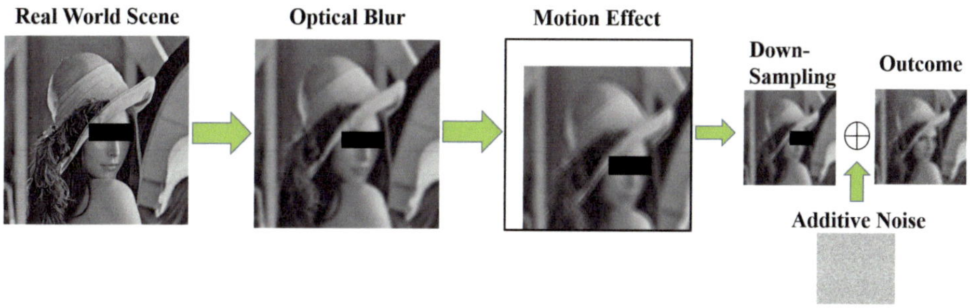

Figure 5.
Conceptual illustration of image degrading effects during conventional camera image acquisition.

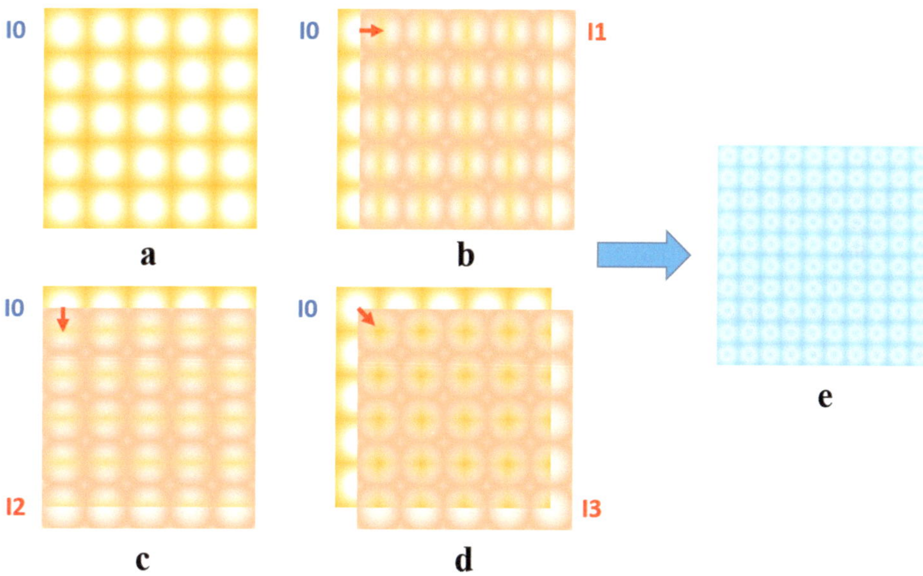

Figure 6.
The illustration shows our superresolution reconstructed image from multiple low resolution frames with sub-pixel shifts. (a) The first x-y plane image I0 with 5 × 5 pixels is set as position 0, as reference. (b) The second image I1 shifts half-pixel to the right of I0. (c) The third image I2 shifts half-pixel to the bottom direction. (d) The last image I3 shifts half-pixel to the right-diagonal direction. (e) Four times pixel resolution (10 × 10) image is obtained after superresolution processing.

superresolution principle [52–54], the resolution improvement comes from the effective sub-pixel information differences among multi-frame low resolution images, as illustrated in **Figure 6**. Without introducing sub-pixel shifts among images, the stationary multi-frame imaging and processing would mainly contribute to minimize temporal noises [48]. For SD-OCT superresolution, the conventional sub-pixel shift now called sub-spot-spacing shift is due to different imaging principles.

Figure 6 illustrates how to apply multi-frame superresolution technique to SD-OCT imaging. A reference 5 × 5 pixel image in one depth layer of a C-scan with no scan spot overlapping is shown in **Figure 6(a)**. As we introduced above, each pixel represents a focused scan beam spot on the sample. To satisfy multi-frame superresolution requirement, we intentionally introduce slight differences in a series of control voltage matrix of scanners, creating a sequence of C-scans with sub-spot-spacing shifts (equivalent to sub-pixel shifts) in the x-y lateral plane, as illustrated in **Figure 6(b)–(d)**. The superresolution processing of the four lateral images but with sub-spot-spacing shifts reconstructs a higher pixel resolution image of that depth layer, exhibited in **Figure 6(e)**.

The image resolution covers two concepts: one is pixel resolution which is equivalent to dots per inch or sampling rate in conventional terminology and the other is spatial resolution which is defined as the smallest discernible detail in an image [55], one example as Rayleigh criterion. **Figure 6** is obviously a result of lateral pixel resolution improvement leading to lateral spatial resolution improvement by the increment of sampling rate. In theory, simply reducing the step size of the sub-spot-spacing shift and increasing non-identical image sets for multi-frame superresolution processing can continuously improve the lateral pixel resolution. However, there is a spatial resolution limit owing to optical diffraction limit, system noises, stability of interference pattern and so on. In other words, when the sampling rate is high enough, further increment would not be helpful to lateral resolution improvement. Thus, finding an effective relationship among the lateral resolution improvement, the sub-spot-spacing shift, and the number of image frames would be critical to identify a desired resolution improvement without unnecessary image frames and associated excess acquisition time. Without particularly indicated, the lateral resolution improvement discussed in this book chapter represents lateral spatial resolution improvement.

Figure 6 illustrates four C-scans having multi-directional 1/2-sub-spot-spacing shifts. For easier explanation, the four shifts are simplified as four blocks in **Figure 7(b)**, showing the shifts directions and space relative to the first non-shift C-scan as reference 0. Mathematically speaking, the three shifts should be represented as (0.5, 0), (0, 0.5), and (0.5, 0.5) in x and y coordinates. Compared with the traditional four-frame shift strategy in **Figure 7(b)**, we experimentally found that more shifts (gray ones) in **Figure 7(c)** and **(d)** in addition to red shifts lead to better image quality. The gray shifts in **Figure 7** provide more information for superresolution processing, suppressing background noises in OCT imaging. By using 1/4-spot-spacing shift as in **Figure 7(d)** red points, the superresolution technique can improve the lateral pixel resolution by 16 times in principle. Similarly, a series 1/8-spot-spacing shift C-scans (not shown) can improve the lateral pixel resolution by 64 times. Simplifying the shift strategy introduced later in the chapter, we name the **Figure 7(c)** as 1/2-spot-spacing shift step and maximum 1/2-spot-shift, and **Figure 7(d)** as 1/4-spot-spacing shift step and maximum 3/4-spot-shift, and so on.

Considering an ideal high quality lateral image $S(x,y)$ degraded by a pure translational motion with space invariant blur and additional noise as $V[x,y]$, one of

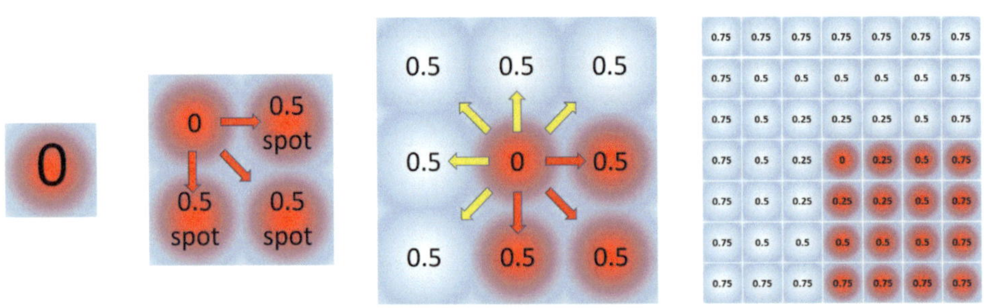

a b c d

Figure 7.
*(a) Take the first C-scan as position 0, for reference. (b) This is a simple illustration of **Figure 6**, showing the traditional shift strategy for four times pixel resolution improvement. (c) Our shift strategy for four times pixel resolution improvement, including additional gray shifts as indicated by yellow arrows, which can provide better output image quality. (d) Similar as (c), we use a smaller 1/4-spot-spacing shift to increase the pixel resolution by 16 times.*

the acquired low resolution lateral image $I[x,y]$ at a selected depth layer in a C-scan is modeled as

$$I[x,y] = [F(H(x,y) \otimes S(x,y))] \downarrow + V[x,y]. \tag{5}$$

Here, F is the motion operator due to sub-spot-spacing shifts among multiple C-scans discussed above. $H(x,y)$ is the PSF of the sample arm optics, blurring the image. $\otimes$ is the convolution operator. $\downarrow$ is the discretizing down-sampling operator due to the sparse scan matrix and finite spot size.

According to the image degradation model in Eq. (5), we can recover the high resolution image $\hat{S}(x,y)$ with a series slight shifted $I[x,y]$'s by mathematical processing. Generally speaking, the recovering processing is minimizing the errors between the model and all the measurement values. We estimate the approximate high resolution image $\hat{S}$ in a minimum L_p norm problem [53] as

$$\hat{S} = \underset{S}{\text{ArgMin}} \left[\sum_{k=1}^{N} \| DF_k HS - I_k \|_p^p \right]. \tag{6}$$

Here, I_k is the kth input low resolution image. F_k is the motion operator for the kth low resolution image. D is the down-sample operator which can be simply determined as $\frac{1}{8}$, $\frac{1}{4}$ or $\frac{1}{2}$ by how many times the sampling rate improvement (such as 8, 4 or 2 times) and the total input frame number captured. H is the optical blur operator or PSF. The noise $V[x,y]$ is an additive term and can be suppressed by multi-frame superresolution processing, which thus is not included in Eq. (6). Besides, we define $G = HS$ as the image S convolved with a PSF, due to the complexity of the deconvolution problem in OCT imaging system. We would solve the deconvolution problem later [44]. Rewriting Eq. (6) [53] we have

$$\hat{G} = \underset{G}{\text{ArgMin}} \left[\sum_{k=1}^{N} \| DF_k G - I_k \|_p^p \right]. \tag{7}$$

Eq. (7) is a minimization of L_p norm problem that can be separated into two steps: reconstruct a non-deconvolved high resolution image $\hat{G}$ from a series of low resolution image frames $\sum_{k=1}^{N} I_k$ and then find a proper PSF to eliminate optical blur H and recover the expected image $\hat{S}$ from $\hat{G}$.

If $p = 1$, it is a L_1 norm problem, or a least-absolute problem. If $p = 2$, it is a L_2 norm problem, or a least-square problem. L_1 norm is robust to outliers but may penalize the high frequency signals. In most OCT applications presented in this chapter, we notice that the background noises are usually temporal noise along with structure related speckle noise without significant outliers. Both the temporal noise and the speckle noise can be suppressed by processing with adjacent pixels [36] and the average of multiple lateral images. Therefore, we applied a kind of L_2 norm called normalized convolution (NC) algorithms introduced by Knutsson et al. [56] and Pham et al. [57] to process the designated shifted images in **Figure 7** to improve the lateral resolution of our SD-OCT system.

We select the NC algorithm [56, 57] instead of other steepest descent algorithms because it considers the relation of a center pixel with neighborhood encompassing N pixels (for example, the radius of 4 pixels). And the final value of each output pixel is optimally solved [58] by adjacent ones, effectively reduce the structure related speckle noise. In experiment section, through shifted C-scans and the NC algorithm, we demonstrated that our superresolution technique can significantly reduce the background noises in final lateral and 3D images. Besides, due to the shift compensation for all low resolution frames, our method avoids ghost patterns observed in output images. Additionally, this kind interpolated method has good tolerance to the incomplete input frames lack of some shifts. For example, even lack of I3 in **Figure 6**, we still can estimate the output image according the neighborhood pixels in incomplete input images.

After the interpolation algorithm, the next step is to find a proper PSF to recover the expected image $\hat{S}$ from $\hat{G}$. There are numerous reports on various deconvolution methods to improve OCT image resolution [45–47]. Lucy-Richardson deconvolution [47, 59, 60] with a proper Gaussian PSF appears to be a widely accepted solution for recovering blurred images,

$$\hat{S}_{m+1}(x,y) = \hat{S}_m(x,y)\left[H(-x,-y)\otimes\frac{\hat{G}(x,y)}{H(x,y)\otimes S_m(x,y)}\right], \tag{8}$$

where $\hat{S}_m(x,y)$ is the estimate of the undistorted image in mth iteration. The deconvolution process starts with $\hat{S}_0(x,y) = \hat{G}(x,y)$. The original input image $\hat{G}(x,y)$ is obtained from Eq. (7). $H(x,y)$ is the lateral PSF of the system. The Gaussian PSF is a common selection [45–47] owing to the focused beam spot lateral profile following a certain Gaussian distribution. However, the spot profile may not keep the circular symmetry for off-axis scanning. Considering the scattering inside a sample, the focused beam may not retain near Gaussian distribution. Thus the blind deconvolution [61, 62] might be a better solution, which uses maximum a posteriori probability (MAP) algorithm to automatically estimate the irregular PSF in the input image and then deblur it, avoiding the limitation of the regular PSF and exhibiting better performance in the final image. In this book chapter, we applied the blind deconvolution method introduced by Krishnan et al. [62]. In theory, the resolution limit of an optical system is determined by diffraction limit [63], which is related to the PSF. Thus, it is possible to break the diffraction limit and further improve the spatial resolution of optical systems through deconvolution with a correct PSF. Although the $\hat{S}$ deconvolved from a Gaussian or estimated PSF would show obvious resolution improvement to $\hat{G}$, these methods may lead to some ringing artifacts and reduce the output image quality. Also, the deconvolution methods are usually sensitive to the noise floor which further restricts their applications, explained later in the experiment Section 3. In this chapter, we thus focus on the first step to reconstruct a high quality image $\hat{G}$, but also provide deconvolved images for readers to compare.

2.4 Estimating the unknown shifts to improve lateral resolution by multi-frame superresolution for *in vivo* imaging

The above superresolution processing is suitable for SD-OCT imaging of static samples such as microstructures where the sub-spot-spacing shifts F_k are intendedly set. For *in vivo* SD-OCT imaging of live tissues such as fingerprint identification, the shifts F_k are unknown due to live body motion and vibration, making the superresolution processing difficult. An effective estimator is critical to accurately estimate the shifts before superresolution processing. We decompose the unknown spatial shifts into two directions: one is in the depth z-axis and the second is in the lateral x-y plane. Herein, the rotational angle motions could be ignored for fingerprint reader.

In the z-axis, the height shifts among multiple C-scans can be corrected by some obvious features, like comparing the top positions of multiple 3D images. While in the lateral x-y plane, without any simple indicators, an advanced shift estimator is desired. To improve the estimation accuracy, we firstly average multiple lateral images along the z-axis to enhance the contrast of key features in the x-y plane. Then a popular image registration algorithm—multi-modal volume registration [64] is applied to estimate the shifts among these averaged lateral images. According to the registration algorithm, we seek to maximize the mutual information between the reference image u and test image v:

$$\hat{T} = \underset{T}{\text{ArgMax}}\, I(u(x), v(T(x))). \tag{9}$$

Here, T is a transformation from the reference image to the test image. $v(T(x))$ is the test image associated with the reference image $u(x)$ after transformed with T. We treat x as a random variable over coordinate locations in u and v. The best transformation $\hat{T}$ can be estimated by algorithms [64–66] to maximize the mutual information I between u and v.

This $\hat{T}$ is considered as the motion operator F_k in Eq. (4) for the kth low resolution *in vivo* lateral image to the reference one. After the approximation of all shifts F_k, the following superresolution processing as described in Section 2.3 would be applied for the lateral resolution improvement. Here, the spatial shifts among multiple C-scans are caused by random body motions and vibrations, and we do not introduce any intended sub-spot-spacing shifts.

2.5 SD-OCT image acquisition and superresolution processing

The SD-OCT image acquisition is a lateral spot scanning image acquisition process where the depth tomographic information in z-axis is obtained intrinsically for

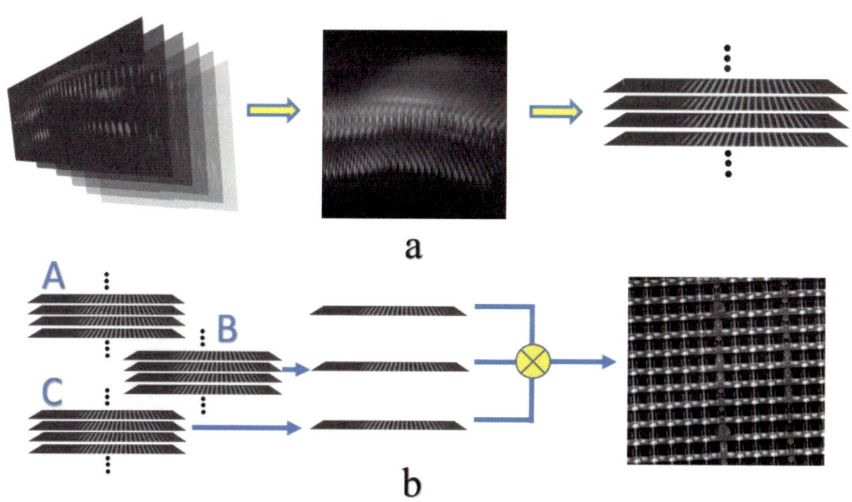

Figure 8.
Extract lateral images from OCT tomography images to reconstruct higher lateral resolution images. (a) Left: The original B-scan tomography images $I(x,z)$ at differenty positions. Middle: Array these B-scans to form a 3D image $I(x,y,z)$. Right: Extract x-y lateral images $I(x, y)$ at different depth z layers to generate a new lateral image stack. (b) Left: A series new lateral shifted image stacks (labeled blue A, B, C) obtained from the processing of (a). Middle: Exact multiple x-y images $I(x, y)$ from the left stacks at the identical depth z. Right: Superresolution processing is applied to these $I(x,y)$ images at the same depth z to generate a high lateral resolution image. Repeat process (b) for all z depth layers can yield a high lateral resolution 3D image stack, not shown.

each scan point. Our superresolution processing is to analyze and improve the lateral resolution in x-y plane. Thus, we need to transfer z-axis information of numerous points to multiple x-y plane layers. First, we perform the SD-OCT C-scan, acquiring multiple B-scan images $I(x,z)$ in the x direction at different y (see **Figure 8(a)**—left). These B-scans can be arrayed in sequence to generate a 3D matrix $I(x,y,z)$ as shown in **Figure 8(a)**—middle. We then retrieve a sequence of 2D x-y images $I(x,y)$ at different depth z as shown in **Figure 8(a)**—right, for later processing. The lateral resolution improvement is to use several x-y images at an identical z position (see **Figure 8(b)**—middle) but from slightly lateral shifted C-scans (A, B, C, etc. from **Figure 8(b)**—left) to perform multi-frame superresolution processing, yielding a higher lateral resolution image $\hat{S}(x,y)$ as in **Figure 8(b)**—right. Repeat the process in **Figure 8(b)** layer by layer for all depth z layers can yield a higher lateral resolution 3D image in the whole space, not shown.

3. Experiments and results

3.1 Lateral resolution, image quality, and efficiency improvement

We compare the performance of our superresolution technique with designated shifts to other traditional methods, such as high density scan and multiple frames averaging, in three aspects: lateral spatial resolution, image quality, and scan time.

1. Lateral spatial resolution: as we mentioned in Section 2, spatial resolution represents the ability to distinguish the smallest discernible detail in the object, such as closed line pairs, which is an important indicator to all imaging systems. A standard negative resolution targets (R3L3S1N—Negative 1951 USAF Test Target, Thorlabs), as partly shown in **Figure 9**, is used to evaluate the resolution improvement. This resolution target provides 10 groups (−2 to +7) with 6 elements per group, offering a highest resolution of 2.19 μm. Considering our beam spot size in **Table 1**, group 4–5 and 6–7 are suitable for resolution testing of our OCT system with 100 and 30 mm focal length lenses, respectively. The resolution (the gap between two lines, the same as the width of 1 line) of group 4–7 is listed in **Table 2**.

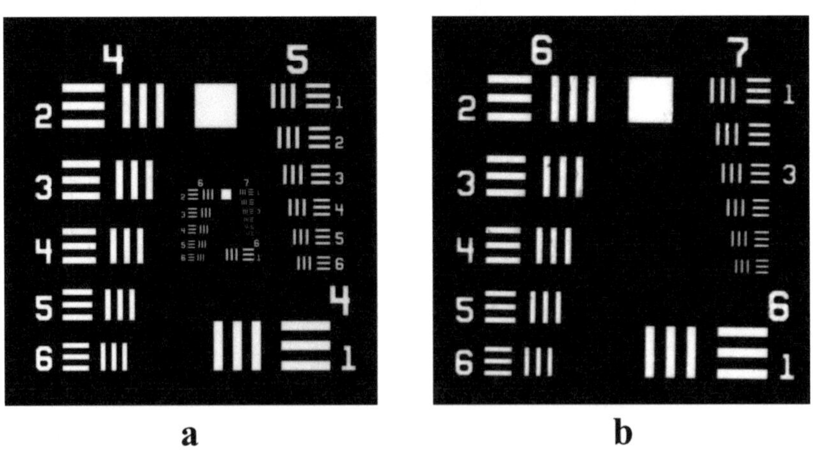

Figure 9.
(a) The image of group 4–7 of the negative 1951 USAF test target is taken by ZEISS SteREO Discovery.V20 microscope. Due to the back illumination, the transparent patterns appear bright white and the chrome portion is dark. (b) The enlarged middle portion of (a) showing details of group 6–7.

Element	Group number							
	4		5		6		7	
	Density of line pairs (lp/mm)	Width of 1 line (μm)	Density of line pairs (lp/mm)	Width of 1 line (μm)	Density of line pairs (lp/mm)	Width of 1 line (μm)	Density of line pairs (lp/mm)	Width of 1 line (μm)
1	16.00	31.25	32.00	15.63	64.00	7.81	128.00	3.91
2	17.96	27.84	35.90	13.92	71.80	6.96	143.70	3.48
3	20.16	24.80	40.30	12.40	80.60	6.20	161.30	3.10
4	22.63	22.10	45.30	11.05	90.50	5.52	181.00	2.76
5	25.40	19.69	50.80	9.84	101.60	4.92	203.20	2.46
6	28.51	17.54	57.00	8.77	114.00	4.38	228.10	2.19

Table 2.
Lookup table of negative 1951 USAF test target.

The resolution target is with a negative clear tone glass pattern. The chrome area appears dark because of blocking the backlight illumination while the transparent patterns are bright. Usually the SD-OCT system is more sensitive to reflectivity enhancement than reduction, and thus a resolution target with sudden reflection reduction is better for judging the resolution limit of the system. Successfully imaging and distinguishing these fine patterns is an effective way to demonstrate both the high lateral resolution and high sensitivity of our technique.

2. Image quality: we take the peak signal-to-noise ratio (PSNR) and the dynamic range (DR) as two indicators to evaluate the image quality improvement. The PSNR definition is given as [36, 67]:

$$PSNR = 20 \log_{10} \frac{Max_{signal}}{STD_{noise}}. \tag{10}$$

Here, STD_{noise} is the standard deviation (STD) of the background noise. Higher PSNR means higher image quality and lower noise. Usually, an acceptable image quality should be with PSNR >20 dB. The DR is defined as [68]:

$$DR = 20 \log_{10} \frac{Max_{signal}}{RMS_{noise}}. \tag{11}$$

Here, RMS_{noise} is root mean square (RMS) of dark noise. Higher DR means we can distinguish more details in both dark and bright areas of an image. For an OCT system, we expect to extract more information of deep layers, imaging weak structure signal from the noise.

3. Scan time: in order to compare the scan time of different methods in a simple way, we take the scan time of 64×64 matrix as unit 1 ($\sim$0.18 s) for reference. Higher density 128×128 scan takes 4 units. Superresolution with 9 shifted low density C-scans of 64×64 takes scan time of 9 units. In experiment, we buffer the scan data and perform the fast Fourier transform subsequently to ensure the shortest scan time. Shorter scan time is very important for *in vivo* 3D imaging avoiding motion errors and artifacts [69]. Even for 3D imaging of static non-biomedical samples, a short scan time would still be needed to

reduce the waiting time and improve the work efficiency, especially in the mass production.

Using a 100 mm focal length lens with a *NA* of 0.015 we performed SD-OCT imaging of the resolution target shown in **Figure 9(a)**. In **Figures 10** and **11**, a set of OCT lateral images are compared, which were acquired by different scan matrixes and processing methods but with the same FOV of $\sim 1 \times 1$ mm^2. All the images were taken in the same experiment with the same focusing condition and light source power. The output images were uniformly set as 8-bit gray TIFF format for comparison.

In **Figure 10**, the scan matrix is given in the first column (such as 64×64) and the corresponding scan time (taking 64×64 scan time as unit 1) in the second column. Column 3 is the OCT lateral image of the resolution target. Column 4 shows the enlarged image of the blue area of the column 3, comparing the barely distinguishable element (the first row) and the indistinguishable element (the last row). Here, we simplify Group i Element j on the resolution target as GiEj. The background noise image of the red region in column 3 is enlarged in column 5 with detailed noise statistics (STD, PSNR, RMS and DR values). The comparisons on lateral resolution, image quality, and scan time in **Figures 10** and **11** are summarized in **Table 3**.

A. Lateral spatial resolution. **Figure 10**(A) is the reference low resolution image with 64×64 scan matrix. Thus, there is no beam spot overlapping like **Figure 3(b)**. We barely see the resolution element in G4E3 which spatial resolution is about 25 μm. Such low resolution is due to the low scan density or undersampling. When increasing the scan matrix to 128×128 (B), 256×256 (C), 512×512 (D), 1024×1024 (E), and 2048×2048 (F) within the same fixed FOV, lateral resolution is obviously improved, indicating the higher scan matrix density in general can contribute to the lateral resolution improvement. However, increasing the scan matrix density from 1024×1024 to 2048×2048, we only observe slight improvement. Further increasing the scan matrix density will not contribute to the lateral resolution but significantly prolong the scan time. From this trend, the maximum resolution is barely seen in 1024×1024 lateral image (E) as G5E4 line width of 11.05 μm which is close to our focused beam spot radius of 10.5 μm.

Except the scan density, further increasing lateral resolution should consider suppressing the background noise. We applied the traditional multi-frame averaging approach to average five of 1024×1024 scanned lateral images, resulting in an improved image in (G) showing visibility of G5E5 of 9.84 μm line width while G5E6 still indistinguishable as the profile in (L) left. Averaging more frames such as 10 would further reduce the noise but cannot improve the lateral solution to G5E6 (not shown here). Also, 10-frame lateral averaging takes too much scan time, unacceptable in a practical OCT 3D imaging.

Compared with high scan density and multi-frame averaging methods, our superresolution processing with designedly shifted low resolution C-scans can effectively improve the lateral resolution. **Figure 10**(H) shows our superresolution processed image with 961 input low resolution shifted C-scans (1/16-spot-spacing step and maximum 15/16-spot-shift). It is a 31×31 shifted scan matrix similar as the 7×7 matrix in **Figure 7(d)**. From the enlarged resolution image in column 4, we can distinguish G5E6 of 8.77 μm line width which is also verified in (L) right. Besides, in order to verify the effectiveness of the superresolution algorithm, we up sampled the 961 input low resolution images to the same image size as (H) by bicubic interpolation, and then averaged them with shift compensation. Although the output image (K) has the same image size of (H), the spatial resolution is terrible, barely observing 12.40 μm line width pattern (G5E3), worse than both the

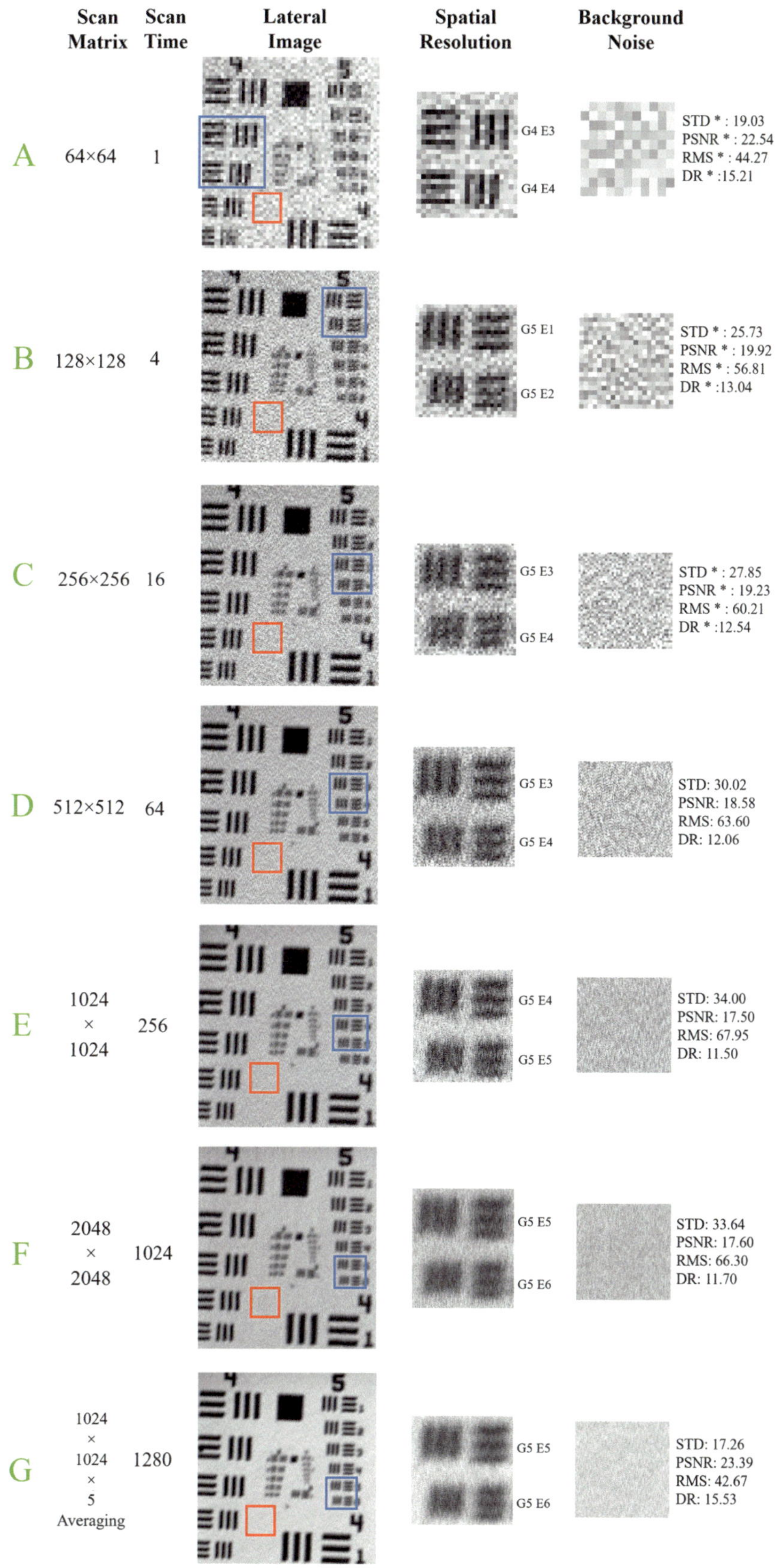
Scan Matrix
Scan Time
Lateral Image
Spatial Resolution
Background Noise
A
64×64
1
G4 E3
G4 E4
STD * : 19.03
PSNR * : 22.54
RMS * : 44.27
DR * :15.21
B
128×128
4
G5 E1
G5 E2
STD * : 25.73
PSNR * : 19.92
RMS * : 56.81
DR * :13.04
C
256×256
16
G5 E3
G5 E4
STD * : 27.85
PSNR * : 19.23
RMS * : 60.21
DR * :12.54
D
512×512
64
G5 E3
G5 E4
STD: 30.02
PSNR: 18.58
RMS: 63.60
DR: 12.06
E
1024 × 1024
256
G5 E4
G5 E5
STD: 34.00
PSNR: 17.50
RMS: 67.95
DR: 11.50
F
2048 × 2048
1024
G5 E5
G5 E6
STD: 33.64
PSNR: 17.60
RMS: 66.30
DR: 11.70
G
1024 × 1024 × 5 Averaging
1280
G5 E5
G5 E6
STD: 17.26
PSNR: 23.39
RMS: 42.67
DR: 15.53

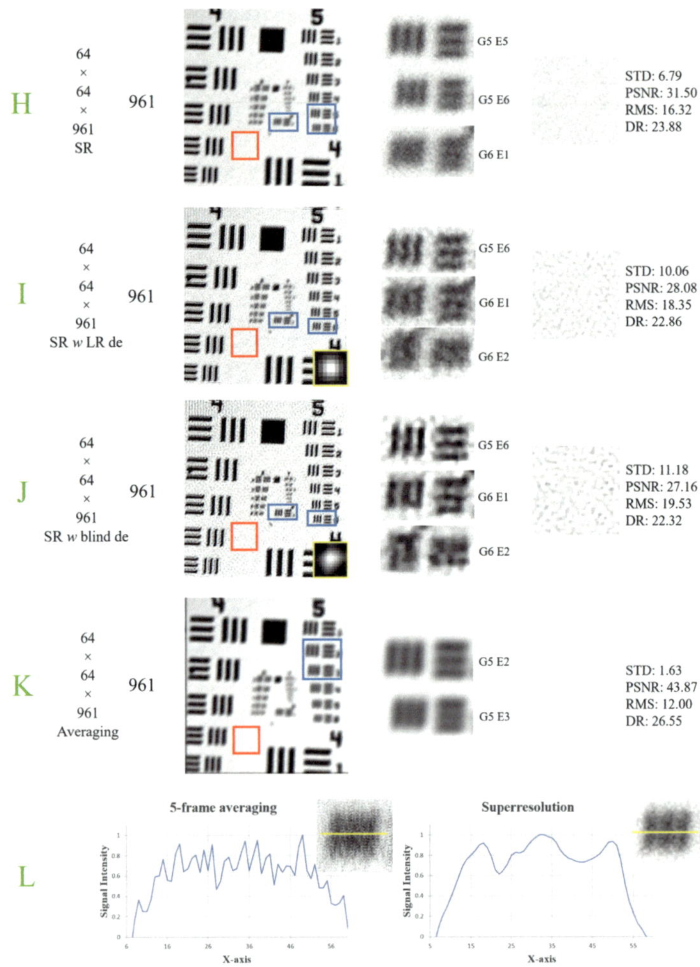

Figure 10.
OCT lateral images of the negative resolution target (group 5–7 region) with the same fixed FOV (~1 × 1 mm²) but different acquisition methods. The scan matrix is shown in column 1 and the corresponding scan time is in column 2. The whole OCT lateral image is exhibited in column 3. The blue square in the image shows distinguishable and indistinguishable elements, enlarged in column 4 for further resolution limit comparison. The background noise in the selected ROI (the red square region in column 3) with STD, PSNR, RMS and DR values is enlarged in column 5. The original low density scan with 64 × 64 non-overlapped spot array is shown in row (A) and set its scan time as reference unit 1. OCT images acquired with different scan matrixes are given in rows (B)–(F). The averaged result of five 1024 × 1024 frames is given in (G). The reconstructed image by superresolution processing from 961 shifted 64 × 64 low resolution input images (1/16-spot-spacing step and maximum 15/16-spot-shift, forming a 31 × 31 shift matrix similar to ***Figure 7(d)****) without deconvolution processing (H) and with Lucy-Richardson deconvolution processing an optimized Gaussian PSF (I) or with the blind deconvolution (J). We average the same 961 low resolution input images with up sampling and shift compensation, shown in (K). All the four images (H)–(K) have the same translation shift parameters and the same output image size. The optimized Gaussian PSF and the estimated PSF by the blind deconvolution are shown at the right bottom of resolution image in (I) and (J). (L) the inverted x-axis horizontal profiles of the yellow lines in the center of patterns G5E6 of (G) and (H), demonstrating the effective resolution improvement in (H). By the visual comparison, the image resolutions in (I) and (J) are obviously better than that (H). *The statistical values are for reference only since much fewer pixels in ROIs of (A)–(C), compared to the pixel numbers in (D)–(K). SR in (H) is short for superresolution. SR w LR de in (I) is short for superresolution with Lucy-Richardson deconvolution. SR w blind de in (J) is short for superresolution with blind deconvolution.*

high density scan and the multi-frame averaging. This comparison demonstrates that the lateral resolution improvement is from both sub-spot-spacing shifted information and the superresolution algorithm, not only more data collection.

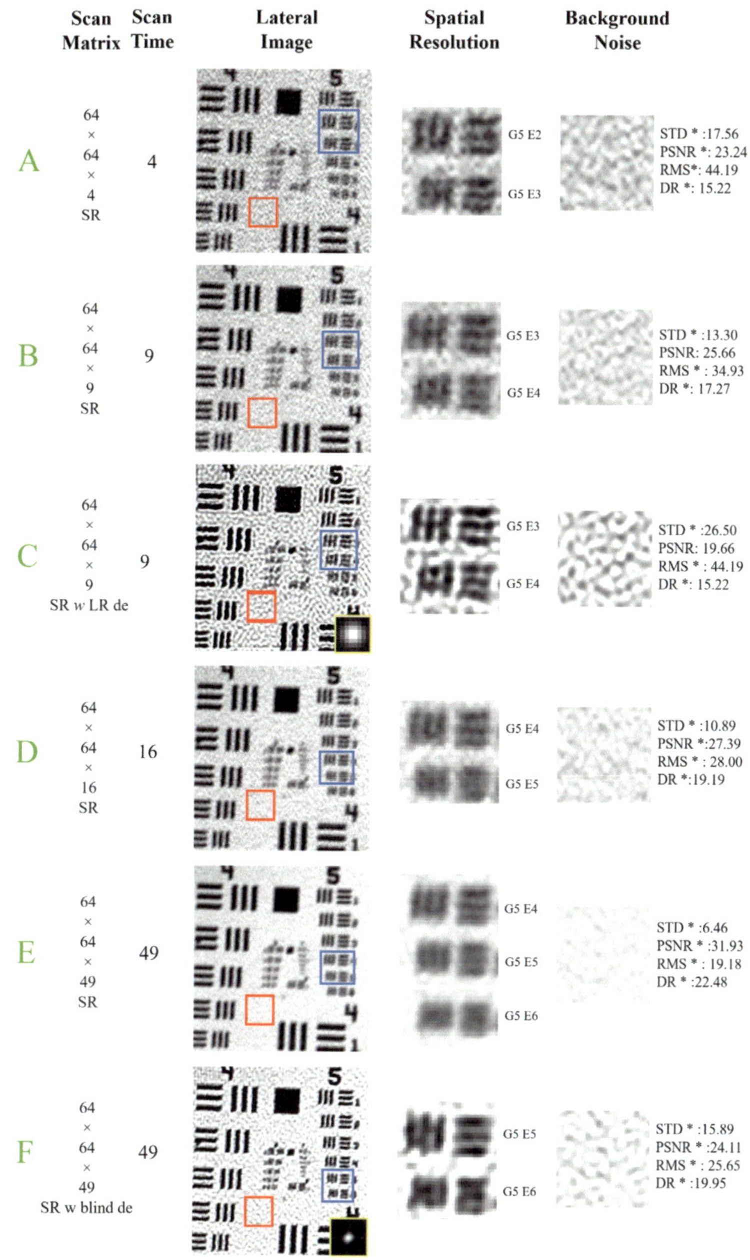

Figure 11.
A list of superresolution processed images with much fewer input C-scans. Superresolution processed images (A) and (B) with scan strategies as in ***Figure 7(b)*** *and* ***(c)****, respectively. Superresolution processed images (D) and (E) with scan strategies as in* ***Figure 7(d)*** *without and with gray shifts, respectively. Image (C) is the Lucy-Richardson deconvolution of image (B) using an optimized Gaussian PSF and the Gaussian PSF is shown at the right bottom of (C). Superresolution with blind deconvolution reconstructed image (F) from the same input images as (E) and the estimated PSF is shown at the right bottom of (F). *The values are for reference only due to low pixel numbers in ROIs. SR in (A)–(E) is short for superresolution. SR* w *LR de in (C) is short for superresolution with Lucy-Richardson deconvolution. SR* w *blind de in (F) is short for superresolution with blind deconvolution.*

After reconstructing the non-deconvolved high resolution image (H) from a series of low resolution images, further lateral resolution improvement should be achieved by Lucy-Richardson deconvolution processing of image (H) with an optimized Gaussian PSF in (I) or by blind deconvolution processing shown in (J) as we discussed in Section 2, both clearly exhibiting G6E1 of 7.81 μm line width without

Scan time[a]	High density scanning					Multi-frame superresolution				
	Scan matrix[b]	Spot spacing	Lateral resolution (μm)	PSNR[c] (dB)	DR[c] (dB)	Low resolution C-scans	Shift	Lateral resolution (μm)	PSNR[c] (dB)	DR[c] (dB)
1	64^2	1	24.80	22.54*	15.21*	1	0	24.80	22.54*	15.21*
4	128^2	1/2	15.63	19.92*	13.04*	4	1/2	13.92	23.24*	15.22*
9	—	—	—	—	—	9	1/2	12.40	25.66*	17.27*
						9de[4]	1/2	11.05–12.40	19.67*	15.24*
16	256^2	1/4	12.40	19.23*	12.54*	16	1/4	11.05	27.39*	19.19*
49	—	—	—	—	—	49	1/4	9.84–11.05	31.93*	22.48*
						49de[d]	1/4	9.84	24.10*	19.95*
64	512^2	1/8	11.05–12.4	18.58	12.06	—	—	—	—	—
225	—	—	—	—	—	225	1/8	9.84	32.58	22.87
						255de[d]	1/8	7.81–8.77	27.40	21.77
256	1024^2	1/16	11.05	17.50	11.50	—	—	—	—	—
961	—	—	—	—	—	961	1/16	8.77	31.50	23.88
						961de[d]	1/16	7.81	27.16	22.32
1024	2048^2	1/32	9.84–11.05	17.60	11.70	—	—	—	—	—
1280	$1024^2 \times 5$	1/16	9.84	23.39	15.53	—	—	—	—	—

The NA of the measurement arm lens is 0.015 with focused beam spot size of about 21 μm. The focal length of the lens is 100 mm.
[a]*The scan time here take the original low density 64 × 64 C-scan without spot overlapping as unit 1 for reference.*
[b]*64^2 means 64 × 64, 128^2 means 128 × 128, and so on.*
[c]*The higher, the better.*
[d]*"de" means superresolution with deconvolution processing shown in **Figures 10**(I), (J) and **11**(C), (F).*
**The values are only for reference.*

Table 3.
Comparison of the lateral resolution and image quality VS scan time by different methods.

additional hardware configuration. Also, the superresolution with deconvolution processing obviously enhances the contrast of the resolution element. All three lines in G5E6 in (I) and (J) are much clearer than in (H), indicating the effectiveness of deconvolution methods. The optimized Gaussian PSF was selected by iteration changing of Gaussian parameters to achieve the best output image. Different from the Lucy-Richardson deconvolution with a manually selected Gaussian PSF, the blind deconvolution can automatically estimate an optimized irregular PSF and thus deblurred the image in (H) with less ringing artifacts, although it still introduces some additional noise to the background. Thus, for the following deconvolution processing, we mainly use the blind deconvolution algorithm. We also attached the Gaussian PSF or estimated irregular PSF at the right bottom of the deconvoluted image.

Compared with original C-scan (A), our superresolution technique improves the lateral resolution from 25 to 8.77 μm (H) (without deconvolution processing) and to 7.81 μm (with deconvolution processing, in (I) and (J)), a factor of ∼3 times improvement. According to the above discussion, we can summarize that for lateral resolution improvement, the superresolution technique with shifted low density C-scans is better than multi-frame averaging of several high density C-scans, which is better than one set simple high density C-scan. The superresolution with deconvolution processing will further improve the lateral resolution.

According to Rayleigh criterion [63], the resolution limit of an optical system is restricted to half of the focused spot size. Our present beam spot radius was measured as ∼10.5 μm, similar to the 9.84 μm line width of G5E5 in **Figure 10**(G). This agrees well with the theory of diffraction limit. We can actually observe the 8.77 μm line width pattern of G5E6 in (H), which is slightly better than the spot radius due to increase of pixel density, reduction of noise, and enhancement of image contrast by our superresolution technique.

The resolution of an optical system is physically restricted by the diffraction limited, or PSF in other words. Dense patterns cannot be distinguished are due to finite spot size blurring. The digital deconvolution processing with a proper PSF can break the diffraction limit for resolution and sharpness improvement. Superresolution processing with Lucy-Richardson deconvolution using an optimized Gaussian PSF in **Figure 10**(I) or with blind deconvolution in (J) can clearly exhibit the G5E6 line width of 8.77 μm with higher image contrast and further show the next group element G6E1 with 7.81 μm line width, both breaking the diffraction limit and significantly improving the lateral resolution.

B. Image quality. Simple high density scan did not change the image quality. Taking the six images in **Figure 10**(A)–(F) as examples, all their PSNRs were almost lower than 20 dB, demonstrating that the increase of scan density did not do any help to the image quality. Actually, with the exactly same focusing condition and light source power, the six images should have very similar quality. Although we see a little better PSNR and DR in (A)–(C), that is due to not enough pixel numbers in region of interest (ROI) which reduces the statistics reliability. Thus, for image quality comparison with other methods, we use scan matrix of 1024 × 1024 in (E) and 2048 × 2048 in (F) as reference.

Through five-frame averaging, the PSNR in (G) is improved to 23.39 dB, better than the value 17.50 dB in (E). A 10-frame averaging can further reach 27.75 dB (not shown) but it is still lower than 30 dB and doubling the scan time of five-frame averaging. The superresolution processed image in (H) can achieve 31.50 dB PSNR, almost doubling the dB values of the high density scan results in (E) and (F). Although all the images in **Figure 10** have the same 8-bit gray range between 0 and 255, we recognize that the superresolution processed image shows better contrast and thus looks brighter. That is because the higher DR and lower background noise in (H), (I) and (J) makes brighter appearance to human eye observation. The DR value of (H) also doubles the values of (E) and (F) in dB unit. The superresolution with deconvolution processed images in (I) and (J) decrease a little in the image quality as compared to (H), because of the increased background noise by the deconvolution processing. Here, we simply summarize the image quality comparison that the superresolution processing is better than the multi-frame averaging which is better than high density scan. The superresolution with deconvolution improves the image resolution but slightly reduces the image quality.

Besides, we noticed that (K) has the best PSNR and DR value among all the images of **Figure 10**, which comes from averaging the 961 up-sampled low resolution images (the same input images as (H)) with shift compensation. The STD value is only 1.63, exhibiting very smooth background without obvious noises. If only focusing on the image quality values, we may be misled that the average of up-sampled images can provide better background noise suppression than the superresolution technique. However, this method penalizes the high frequency signal, resulting in a poor resolution of 12.4 μm, even worse than the high density scan in (E), which is not an acceptable method.

C. Scan time. From column 2 of **Figure 10**(F)–(J), it is easy to summarize that the scan time of the superresolution technique is faster than both the high density scan and the multi-frame averaging. Superresolution processing provides much better image resolution and quality with less scan time. **Figure 10** compares the resolution limit of different methods and thus takes long scan time. For example, the present scan time of **Figure 10**(A) for FOV of $\sim 1 \times 1$ mm^2 takes 0.18 s while that of (F), (G), and (H) take 3.41, 4.27, and 3.2 min, respectively. If enlarging the FOV to $\sim 3 \times 3$ mm^2 area and keeping the same scan density of (F), (G), and (H), these methods would take 30.7, 38.4, and 28.8 min scan time (excluding fast Fourier transform calculation), too long for many applications. In practice, we need to consider acceptable scan time for *in vivo* imaging and the effectiveness of the experiments.

To reduce the scan time, we compare a list of superresolution processed images in **Figure 11** with much fewer input C-scans than **Figure 10**(H). Also, there are two different shift strategies applied in this experiment similar as **Figure 7(b)** and **(c)** to demonstrate the additional gray shifts in **Figure 7(c)** are needed for higher lateral resolution and image quality. Although the red shifts are enough for sampling rate improvement by superresolution processing, those additional gray shifts could contribute to image noise reduction, the lateral spatial resolution and overall image quality improvement. In **Figure 11**(A), the pattern G5E3 is indistinguishable, processed with **Figure 7(b)** scan strategy. While with more shifts as **Figure 7(c)** strategy, the pattern of G5E3 in **Figure 11**(B) is clearly visible and we can further partly distinguish the G5E4 pattern. Similarly, the G5E5 in **Figure 11**(D) is not obvious with red shifts only in **Figure 7(d)**. After including the additional gray shift patterns, the G5E5 pattern in **Figure 11**(E) becomes visible. Thus, these gray shifts can effectively improve the lateral resolution as well as reduce the background noise by about 20–70% in STD and RMS, overcoming the reconstructed image with red shifts only.

The superresolution processing with Lucy-Richardson or blind deconvolution has demonstrated its contribution to the resolution improvement again, shown in **Figure 11**(C) and (F) as compared to (B) and (E), respectively. The deconvolution also introduces some degradation to image quality, increasing the background noise similar as in **Figure 10**(I) and (J). It is important to note that the superresolution with deconvolution does not spend any extra scan time.

Comparing with **Figure 10**(B)–(E), the results in **Figure 11** clearly show the advantage of our multi-frame superresolution processing with less scan time while offering much better lateral resolution and image quality. Reducing from 961 to 49 shifted C-scans, it only takes 9.8 s to see the 9.84 μm line width pattern in **Figure 11**(F), while the 1024 × 1024 high density scan in **Figure 10**(E) takes about 51 seconds to barely observe the 11.05 μm line width pattern with lower image quality. Similarly, **Figure 11**(A)–(E) provide higher lateral resolution and better image quality with shorter scan time than **Figure 10**(B)–(E). Clearly, our superresolution technique has demonstrated its superior performance in lateral resolution and image quality improvement with shorter scan time.

Based on the above experiments, the lateral resolution and image quality vs. scan time are summarized in **Table 3**. Obviously, the multi-frame superresolution technique can achieve much better lateral resolution and image quality with less scan time than high density scanning and multi-frame averaging.

Except the experiment with 100 mm focal length lens above, we also checked the performance of a 30 mm focal length lens, which focuses the collimated beam to the diameter of ~6 μm, very suitable to image the patterns group 6–7 in the resolution target of **Figure 9(b)**. **Figure 12**(A)–(D) exhibits the original, the high density scanned, the average of multiple high density scans, and our superresolution with deconvolution processed images, respectively. The original low density scan (64 × 64) cannot distinguish any pattern, except the G6E1 with line width of 7.81 μm in **Figure 12**(A). With extremely higher scan density of 2048 × 2048 (taking 1024 scan time units) or averaging of five 1024 × 1024 scanned images (1280 time units), the 3.10 μm (G7E3) and the 2.76 μm (G7E4) become barely visible as shown in **Figure 12**(B) and (C). After the multi-frame superresolution with blind deconvolution processing of 961 shifted low resolution images (similar as **Figure 12**(A), with 1/16-spot-spacing shift step and maximum 15/16-spot-shift), we can see the 2.19 μm patterns (G7E6) as in **Figure 12**(D). The lateral resolution has been significantly improved from 7.81 μm (the original sparse scan in **Figure 12**(A)) to 2.47 μm (superresolution processing without deconvolution, not shown) and to 2.19 μm (superresolution processing with blind deconvolution in **Figure 12**(D)), about 3–3.5 times improvement. Compared with other methods like the high density scan and the multi-frame averaging, our superresolution technique exhibits superior advantage in lateral resolution improvement again. Our technique also shows the apparently better image quality than other methods: PSNR and DR of 103.7 and 137.9% (without deconvolution) and 65.2 and 106.3% (with deconvolution, **Figure 12**(D)) higher than high density scan (**Figure 12**(B)) in dB unit; PSNR and DR of 50.9 and 60.5% (without deconvolution) and 22.4 and 39.2% (with deconvolution, **Figure 12**(D)) higher than the multi-frame averaging (**Figure 12**(C)) in dB unit. Similar as the experiment of using 100 mm focal length lens, the use of 30 mm focal length lens demonstrates again that our superresolution technique can offer higher lateral resolution and better image quality with less scan time than the high density C-scan and the multi-frame averaging method.

The present Lucy-Richardson deconvolution with a Gaussian PSF or the blind deconvolution with an estimated PSF however have some problems: Although the deconvolution effectively improves the lateral resolution, it introduces some artifacts in **Figures 10**(I), (J), **11**(C), (F) and **12**(D), which may not be acceptable for some critical applications. The artifacts are from both imperfect PSF selection and the discrete Fourier transform. And they cannot be avoided in the advanced blind deconvolution.

To our observation, the deconvolution methods are sensitive to the noise level. If background noise is as low as **Figure 10**(K), the deconvolution processing will not introduce obvious artifacts (not shown here, referring to our previous work [11]). However, the method in **Figure 10**(K) is harmful to the spatial resolution.

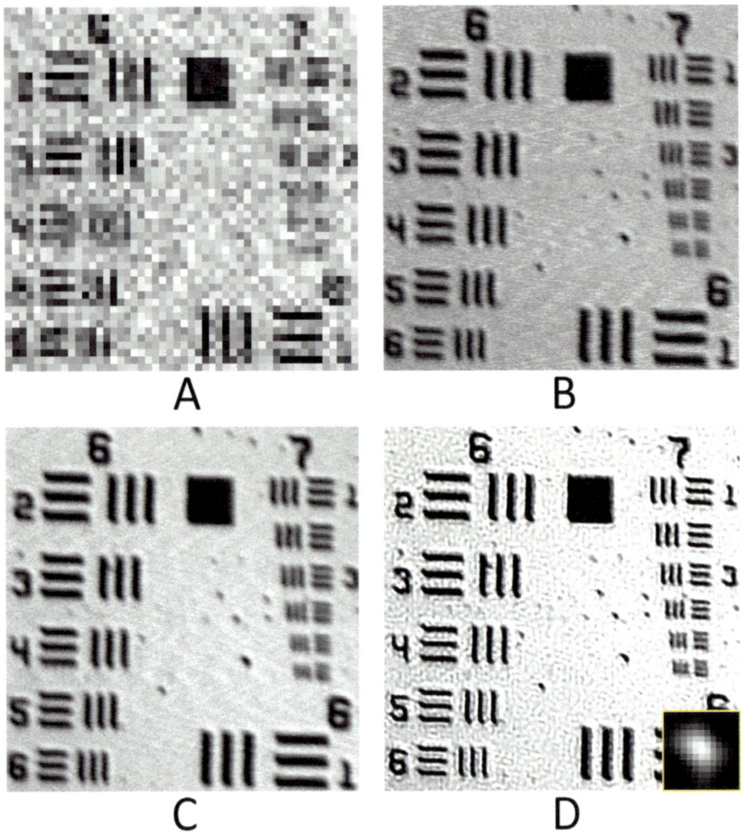

Figure 12.
The OCT lateral images of the resolution target in ***Figure 9(b)*** *were taken by a 30 mm focal length achromatic lens using (A) 64 × 64 scan matrix, (B) 2048 × 2048 scan matrix, (C) 1024 × 1024 scan matrix with 5 frame averaging, and (D) the superresolution and blind deconvolution processed image from 961 shifted 64 × 64 low resolution images with 1/16-spot-spacing step and maximum 15/16-spot-shift. The estimated PSF is attached at the right bottom of (D). All the four images have the same FOV of about 250 × 250 μm².*

Practically, it is difficult to obtain a penetrated lateral image with so smooth background as well as maintaining high resolution due to various scattering mediums in the samples.

When a focused beam penetrating into a sample, the scattering would alter the cross-section profile of the beam. The optimized lateral PSF thus may be different in different samples and at different depth layers [45]. Even with advanced blind deconvolution, the ground true PSF [70] of the system at that depth layer is still difficult to find. We also could notice that the estimated PSFs in **Figures 10**(J), **11** (F) and **12**(D) are different.

Considering the above issues, we would not apply the deconvolution processing to the following OCT experiments of functional samples. However, it is important to point out that the superresolution technique with deconvolution processing can break the diffraction limit, improve the sampling rate and suppress the background noise together to significantly improve the lateral resolution and image quality.

3.2 Improved lateral resolution imaging of microstructure samples

Thus far, we have successfully demonstrated the effective lateral resolution and image quality improvement by the multi-frame superresolution processing with shifted low resolution C-scans. This processing can offer better image quality with less scan time than high density C-scan images and is especially suitable for imaging micron scale fine structures [19, 37, 71–74].

We examined 3D imaging of a microstructure sample in **Figure 13**, in which the particle size is about 3 μm. The two left images of **Figure 13(A)** and **(B)** are the original sparse scan lateral SVP images of the same sample using 30 and 19 mm focal length lens with 1300 × 1300 and 500 × 500 μm^2 lateral FOV, respectively. Even with 19 mm focal length lens and ~4 μm focused spot size, the microstructures are still invisible. After superresolution processing of 225 low resolution shifted frames with 1/8-spot-spacing step and maximum 7/8-spot-shift (using 15 × 15 shift matrix with arrangement similar to **Figure 7(d)**), we are able to observe clearly those microstructures and wavy surface caused by the imperfect fabrication in exposure and developing. This wavy surface is difficult to be seen in microscope imaging without topographic imaging capability. As our previous report, the multiple-frame superresolution processing can improve the lateral resolution of our SD-OCT with 19 mm focal length lens by ~3 times, achieving 1–2 μm [37]. Although 19 mm focal

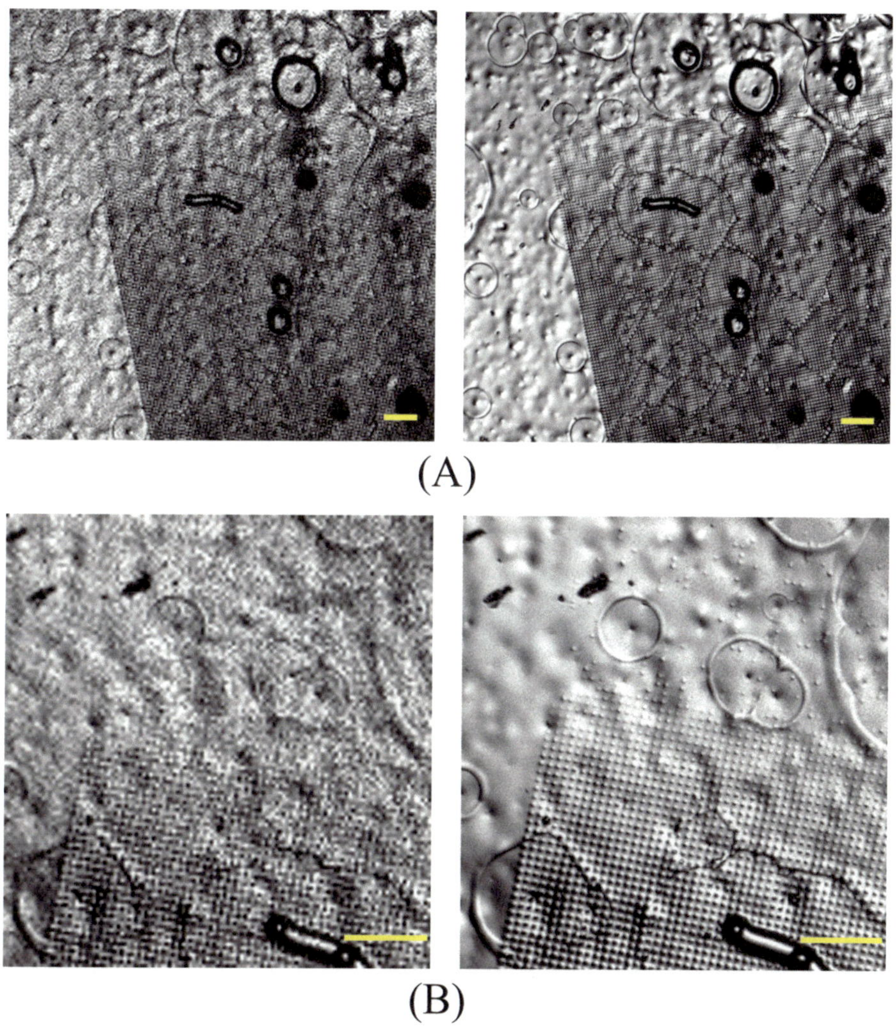

Figure 13.
Lateral SVP imaging of a microstructure sample without (left) and with (right) multi-frame superresolution processing. (A) 30 mm and (B) 19 mm focal length lens are used in the sample arm of our SD-OCT system. Except the ability of observation of ~3 μm particles, the right images show wavy surface with various imperfect fabrications, which is difficult to be observed in microscopy imaging. The scale bars in (A) and (B) are 100 μm.

length lens can provide better image resolution than 30 mm focal length lens due to smaller focal spot, it sacrifices the lateral FOV and axial *DOF* of the system. This trade-off should be considered when imaging different samples. In this experiment, the superresolution enhanced 30 mm focal lens system has provided good enough resolution ability to exhibit the details of the sample.

Except for better human vision, the lateral resolution and image quality improvements further benefit various machine vision algorithms, providing more details for feature detection. Our previous work has reported the superresolution assisted image stitching for achieving an ultra-wide lateral FOV. Taking **Figure 14** as an example, we scanned a multi-layer microfluidic sample by the high density scan and our multi-frame superresolution with shifted C-scans introduced above. All the structures are visible in **Figure 14(A)** left, however with a lot of speckle noises. Applying the advanced SURF [75] feature detection algorithm to the left two adjacent SVP images, there are no correct feature pairs found between them. And the incorrect matching information fails the following image stitching, overlapping two left images as **Figure 14(A)** right. Actually, there is only 30% shared region for the left two images. This failure is because most machine vision algorithms are not robust to periodic structures and noisy background. After superresolution processing, the image quality is significantly improved as in **Figure 14(B)** left, although with the same pixel resolution. The improved images offer much more correct feature pairs, supporting the following image stitching algorithm to reconstruct a wide lateral FOV image successfully at right. This comparison demonstrates the superresolution technique would be an effective pre-processing for subsequent machine vision algorithms.

As we discussed in Section 2.1, each lens has its lateral FOV limitation due to Petzval field curvature. For example, $1400 \times 1400\ \mu m^2$ optimized lateral FOV for 30 mm focal length lens guarantees the overall high resolution for the whole C-scan region. However, this lateral FOV is obviously not enough to image a large sample with centimeter scale sizes. To overcome this drawback, we scan 6 nearby partial overlapped regions of a microstructure sample by a 30 mm focal length lens. Each local C-scan covers a FOV of $1300 \times 1300\ \mu m^2$ and is enhanced by the

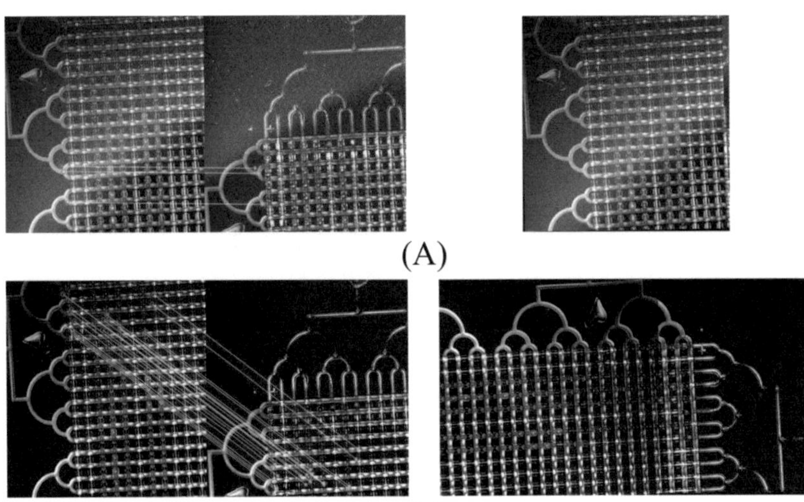

(A)

(B)

Figure 14.
(A) Failed image stitching from two noisy images. (B) The superresolution processed images provide more correct matched feature pairs and successfully help the subsequent stitching algorithm to reconstruct a wide FOV image.

superresolution processing. One of the SVP images is shown in **Figure 13 (A)** right. After repeating the image stitching layer by layer introduced in our previous work [37, 38], we generated a 3.2 × 2.3 mm^2 wide FOV seamless 3D image with high lateral resolution, as shown in **Figure 15(A)**. Wide FOV images at three selected depth layers are shown in **Figure 15**(B). If enlarge the selected two B-scans (positions of the two arrows in the top view) in **Figure 15(C)**, all adjacent parts are also stitched very well without any discontinuities. The details of the image stitching are given in our papers [37, 38].

Again, we stitched 10 close-by C-scans with 500 × 500 μm^2 FOV, imaged by a 19 mm focal length lens, to reproduce a 2.10 × 1.15 mm^2 wide FOV 3D image in **Figure 16**. Due to short focal length lens, this figure stitched by more C-scans only covers 1/3 area of **Figure 15**, although with higher lateral resolution of <2 μm

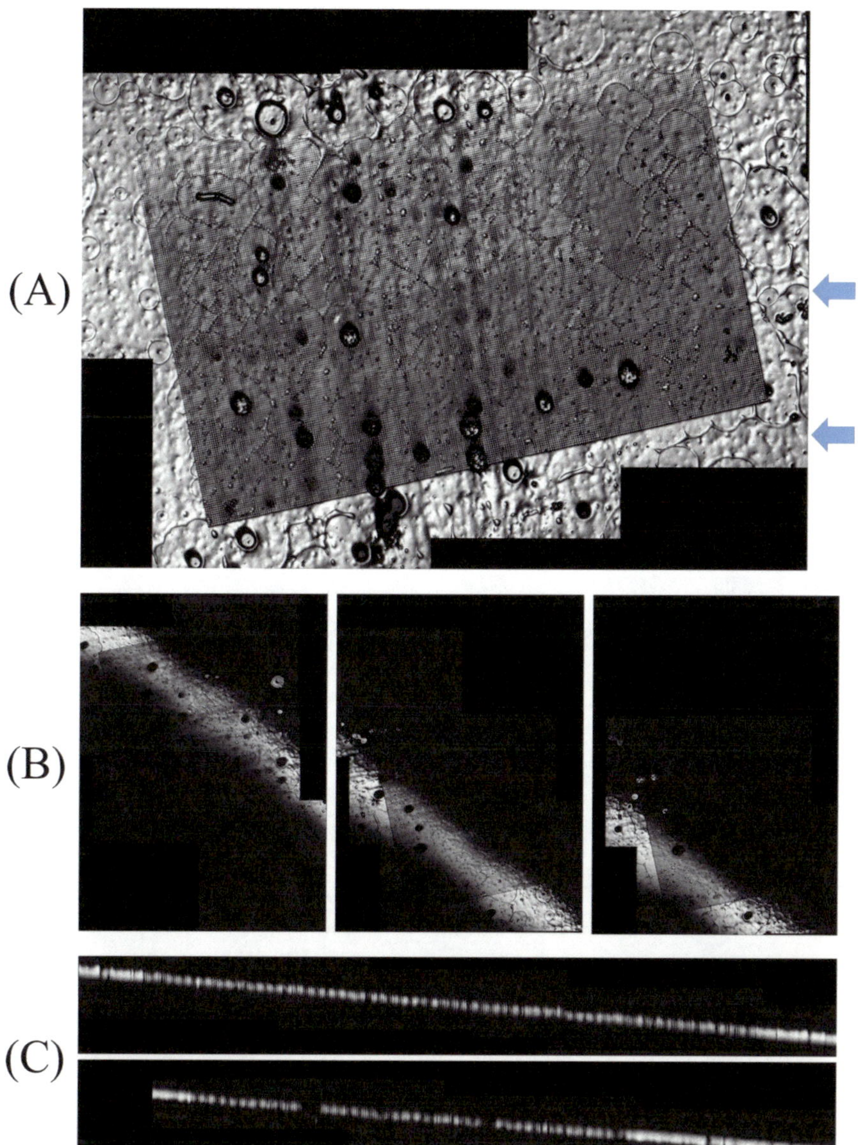

Figure 15.
(A) Top view of a 3.2 × 2.3 mm^2 microstructure image stitched by six 1300 × 1300 μm^2 images. (B) Selected three depth layers in the stitched high lateral resolution 3D image. (C) Two B-scans in the stitched 3D image are selected by the arrows in (A).

[37]. The top view, selected layers, and selected B-scans are exhibited in **Figure 16(A)–(C)**, respectively. The wide FOV images could be enlarged for stitching performance and image quality checking by readers. In principle, there is no limitation on lateral FOV enlargement by this image stitching technique while maintaining the needed high lateral resolution SD-OCT imaging by the multi-frame superresolution processing. While, for fully review the microstructure sample, the 19 mm lens need to image more than 30 adjacent regions due to small lateral FOV, spending at least 5 times more scan time than using a 30 mm focal length lens, thus it is only suitable for ultra-high lateral resolution imaging.

3.3 Improved lateral resolution imaging of *in vivo* 3D fingerprint

The previous section has successfully demonstrated the superresolution processing enhanced 3D imaging for static samples. Actually, this quick and high quality 3D imaging technique is very suitable for time sensitive security applications such as *in vivo* 3D fingerprint identification. The traditional high density scan spends long time and easily leads to motion errors during the scanning. Using our

(A)
(B)
(C)
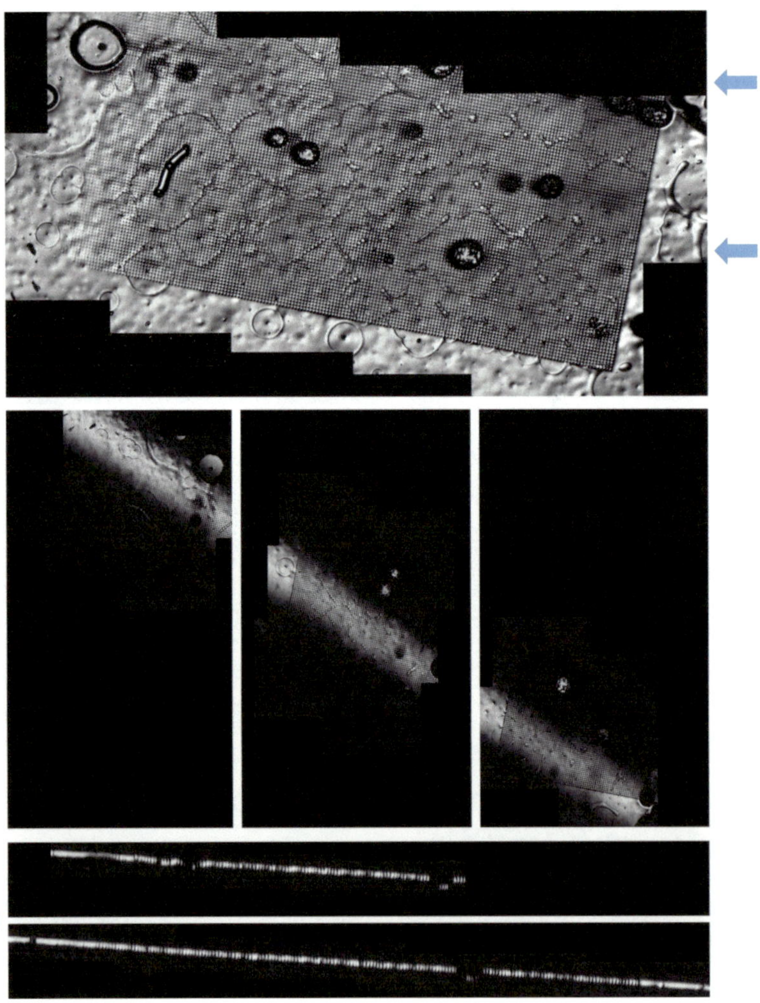

Figure 16.
(A) Top view of a 2.10 × 1.15 mm² microstructure image stitched by ten 500 × 500 μm² images. (B) Selected three depth layers in the stitched high lateral resolution 3D image. (C) Two B-scans in the stitched 3D image are selected by the arrows in (A).

sparse scan method, the SD-OCT only takes 2.8 s to acquire one 256 × 256 C-scan (excluding fast Fourier transform processing time), fast enough to avoid most motion errors within one C-scan cycle, assisted by a finger holder to reduce the potential body motions and vibrations. The *in vivo* unintended tissue movements lead to unknown spatial shifts among multiple C-scans. In order to apply the multi-frame superresolution technique to a series of *in vivo* sparse C-scans, the unknown shifts F_k should be solved first. As we discussed in Section 2, we decompose these unknown spatial shifts into two directions: the depth direction z and the *en-face* lateral plane x-y. The z-axis differences of two C-scans could be estimated by comparing their top positions. For more complex lateral intensity distribution, we utilize the effective multi-modal volume registration [64] to estimate the shift amounts in x- and y-axis for each two SVP images, which provides better lateral details. We also overlap the test image and the reference one with the shift compensation to double check the correctness of estimated lateral shifts. After collecting the x-, y-, z-position shifts information which produces best overlapping quality, the multi-frame superresolution processing is then performed layer-by-layer to improve the lateral resolution and reconstruct a high quality 3D image. The details of the estimation performance and overlap quality are given in our published work [11].

As discussed above, OCT has great potential in security applications, such as *in vivo* 3D fingerprint reader. Currently, fingerprint identification has been a dominant biometry technique, occupying about two-thirds of the biometry identification market [76]. Conventional optical or capacitive acquisitions of fingerprints can only capture a two-dimensional (2D) image of the surface, which have lots of

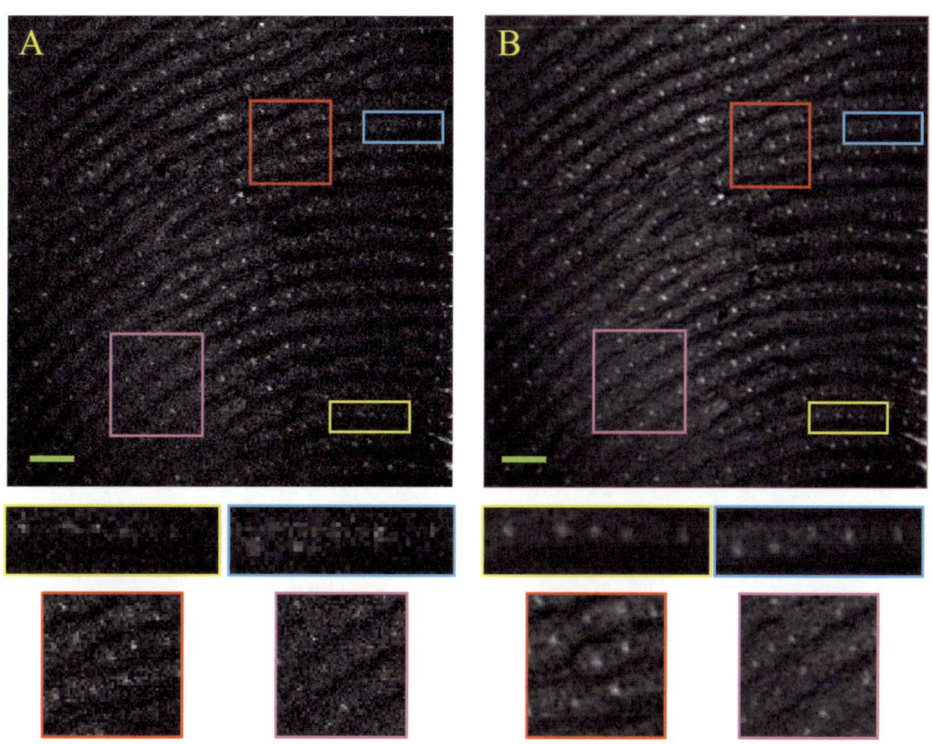

Figure 17.
(A) One original low resolution SVP image of the eccrine sweat glands layer. Below are the enlarged images of selected region with the yellow, blue, red and pink colors. (B) the superresolution processed image showing the improvement in the enlarged local images. The scale bars in (A) and (B) are 500 μm. 100 mm focal length lens is used here.

limitations like pressure dependent skin distortions, skin damages, and wet or fuzzy fingerprints. More seriously, the traditional 2D fingerprint acquisition and analysis are not robust to detect fake fingerprint on spoofing attacks and identity thefts. Our superresolution enhanced SD-OCT could provide high quality 3D image to overcome 2D fingerprint reader. To demonstrate this idea, we examine *in vivo* 3D fingerprint imaging of a thumb (a 33-year-old male volunteer) to show the advantages. Successful imaging of subsurface eccrine sweat glands can serve as a good indictor to the SD-OCT image resolution and effectively defense fake fingerprint attacks which do not have these internal glands. **Figure 17** shows two SVP images (covering about 5×5 mm^2) of the eccrine sweat glands layer, which is the gap between the external and internal fingerprint layers, illustrated in **Figure 18**. The eccrine glands grow under the dermis and open out through the sweat pores on the surface. From the top view of the scanned fingerprint, these glands should appear as the dot style distribution through the whole region. However, due to the low resolution, the SVP image of the original sparse C-scan could not show the eccrine sweat glands distribution clearly. The enlarged yellow and blue local regions in **Figure 17(A)** only barely exhibit some brighter pixels, which cannot be distinguished from the background noise. After superresolution processing with 10 of such C-scans, the reconstructed eccrine sweat glands layer shows much higher lateral resolution and image quality. For example, the five gland spots in yellow and blue selected regions of **Figure 17(B)** can be clearly observed. We are also able to see the low contrast internal structures in the red and pink selected regions of **Figure 17(B)** which however cannot be imaged well in the original C-scans like image in **Figure 17(A)**.

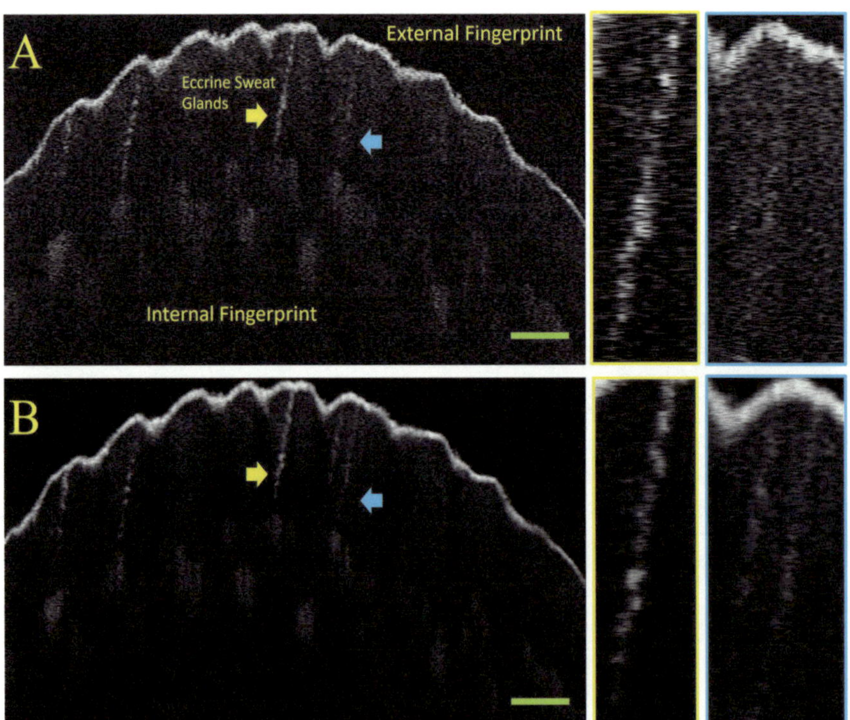

Figure 18.
(A) An original sparse B-scan covered about 5 mm scan width on a thumb. The yellow and blue rectangles are the enlarged areas pointed by the yellow and blue arrows, respectively. (B) One B-scan image extracted from the superresolution processed 3D image, at the same position of (A). The scale bars in (A) and (B) are 500 μm. 100 mm focal length lens was used.

The layer by layer superresolution processing also improves the B-scan image quality. **Figure 18(A)** shows an original low resolution fingerprint B-scan image in the same C-scan of **Figure 17(A)**. We only observe the external fingerprint pattern but with very blurred images of the eccrine sweat glands and the internal fingerprint structures. The two right side images are enlarged areas pointed by the yellow and blue arrows. The yellow rectangle image shows a blurred eccrine sweat gland but we cannot distinguish the helical structure. The blue square image does not exhibit any eccrine sweat glands. After the same superresolution processing as the **Figure 17(B)**, we extracted one B-scan image shown in **Figure 18(B)** from the final high quality 3D image (**Figure 19(A)**) at the same position of **Figure 18(A)**. In **Figure 18(B)**, the helical structure of the eccrine sweat gland marked by the yellow arrow is clearly visible and enlarged at the right side. The three eccrine sweat glands have different intensity because their centers are not in the same B-scan plane. The superresolution processed B-scan exhibits excellent image quality with 49.9% PSNR and 50.6% DR improvement in dB unit. The improvement from **Figure 18 (A)** and (B) can be clearly visualized. After separating the multi-layer fingerprint 3D image (**Figure 19(A)**) into three layers: external fingerprint layer, eccrine sweat glands layer and internal fingerprint layer, the curved layer images are shown in **Figure 19(B)–(D)**, respectively. The distribution of eccrine sweat glands in the whole scan area are beautifully displayed in **Figure 19(C)**. The application of

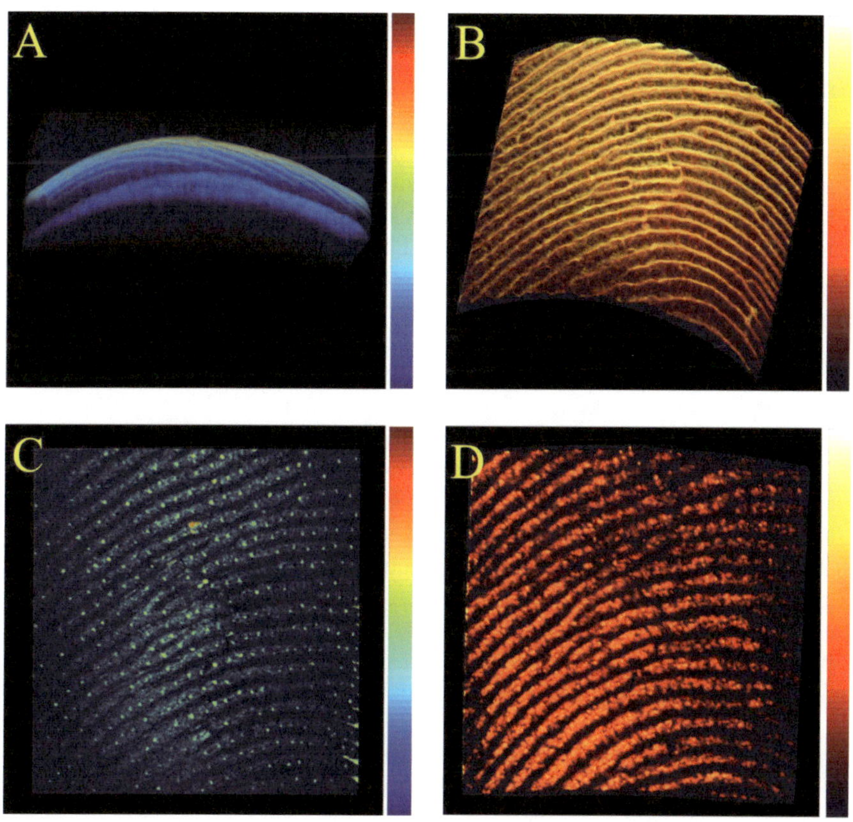

Figure 19.
Superresolution enhanced SD-OCT in vivo *3D imaging of a male thumb fingerprint. (A) A 3D side view shows the multi-layers of a fingerprint. (B) 3D imaging of the external fingerprint layer, processed by our superresolution technique. (C) A 3D top view of the eccrine sweat glands layer after our superresolution processing. Each spot is an eccrine sweat gland. (D) A 3D top view of the internal fingerprint layer improved by superresolution processing, showing the same structure as the external fingerprint. 100 mm focal length lens was used here.*

colormap brings the gland distribution clearer than the gray scale mapping in **Figure 17**. The 3D fingerprint structure is shown in both **Figure 19(B)** and **(D)**. Our superresolution enhanced SD-OCT successfully reconstructs the high quality *in vivo* 3D subsurface fingerprint image. According to other reports, the surface external fingerprint is actually a replicate of the 300 μm lower internal fingerprint structure (the primary ridges) [77]. The high quality imaging of the internal fingerprint with the same features as the surface could be a significantly improved fingerprint identification technique, benefitting from existing large fingerprint database and avoiding the heavy database rebuilding work for other biometric techniques such as iris scanning [78] and face recognition [79], as well as effectively defending against fake fingerprints without such inner structures.

4. Conclusion

In conclusion, a high lateral resolution and high image quality SD-OCT 3D imaging has been achieved by the multi-frame superresolution technique, with shorter scan time than traditional methods. Through adjusting the matrix of control voltages to the galvanometer scanners, we intendedly introduce designed sub-spot-spacing shifts to low resolution C-scans for static sample imaging. After the multi-frame superresolution processing of these shifted C-scan images, about 3 times lateral resolution improvement has been demonstrated by imaging a standard resolution target, from 25 to 7.81 μm and from 7.81 to 2.19 μm with sample arm lens *NA* of 0.015 and 0.05, respectively. Significant background noise reduction and image quality improvement without sacrificing the axial *DOF* and lateral FOV have also been attained. Moreover, the improved lateral resolution and image quality could further benefit various machine vision algorithms sensitive to the noise, providing more features. In combination with our previous work, an ultra-wide lateral FOV and high image resolution and quality OCT has been implemented for static non-medical applications, such as imaging a large microstructure sample.

We present that Lucy-Richardson deconvolution with an optimized Gaussian PSF and the advanced blind deconvolution may potentially break the diffraction limit to further improve the lateral resolution of OCT systems. Although the PSF is highly dependent on samples and depth layers as well as the deconvolutions are sensitive to noise levels, we show the conceptual significance of our superresolution with the following deconvolution in lateral resolution improvement.

For *in vivo* imaging of biometry identification, due to the concern of live body unintended vibration, the multi-volume registration algorithm is used to estimate translational shifts in *x-y* plane without introducing sub-spot-spacing shifts. Then the same multi-frame superresolution processing with the estimated shifts successfully improve the lateral resolution for *in vivo* imaging. The *in vivo* layered 2D lateral images, B-scan tomography images and 3D images of a live fingerprint have shown remarkable lateral resolution and image quality improvement, compared to original C-scan images. The high quality imaging of internal fingerprint and the eccrine sweat glands could effectively defend fake fingerprint on spoofing attacks and identity thefts in important security applications.

Although the present study depends on a SD-OCT system, the superresolution technique is able to work with other scan based OCT imaging system including time domain OCT and swept source OCT, benefiting various medical and non-medical OCT imaging applications.

Acknowledgements

We thank New Span Opto-Technology for providing the SD-OCT system.

Conflict of interest

The authors declare no conflict of interest.

Author details

Kai Shen, Hui Lu, Sarfaraz Baig and Michael R. Wang*
Department of Electrical and Computer Engineering, University of Miami, Miami, Florida, USA

*Address all correspondence to: mwang@miami.edu

References

[1] Huang D, Swanson E, Lin C, Schuman J, Stinson W, Chang W, et al. Optical coherence tomography. Science. 1991;**254**(5035):1178-1181. DOI: 10.1126/science.1957169

[2] Drexler W, Fujimoto JG. Optical Coherence Tomography: Technology and Applications. Heidelberg: Springer Science & Business Media; 2008. DOI: 10.1007/978-3-540-77550-8

[3] Babalola O, Mamalis A, Lev-Tov H, Jagdeo J. Optical coherence tomography (OCT) of collagen in normal skin and skin fibrosis. Archives of Dermatological Research. 2014;**306**(1):1-9. DOI: 10.1007/s00403-013-1417-7

[4] Dalimier E, Salomon D. Full-field optical coherence tomography: A new technology for 3D high-resolution skin imaging. Dermatology. 2012;**224**(1): 84-92. DOI: 10.1159/000337423

[5] Gerger A, Koller S, Weger W, Richtig E, Kerl H, Samonigg H, et al. Sensitivity and specificity of confocal laser-scanning microscopy for *in vivo* diagnosis of malignant skin tumors. Cancer. 2006;**107**(1):193-200. DOI: 10.1002/cncr.21910

[6] Crisan M, Crisan D, Sannino G, Lupsor M, Badea R, Amzica F. Ultrasonographic staging of cutaneous malignant tumors: An ultrasonographic depth index. Archives of Dermatological Research. 2013;**305**(4):305-313. DOI: 10.1007/s00403-013-1321-1

[7] Hee MR, Puliafito CA, Duker JS, Reichel E, Coker JG, Wilkins JR, et al. Topography of diabetic macular edema with optical coherence tomography. Ophthalmology. 1998;**105**(2):360-370. DOI: 10.1016/S0161-6420(98)93601-6

[8] Nassif NA, Cense B, Park BH, Pierce MC, Yun SH, Bouma BE, et al. *In vivo* high-resolution video-rate spectral-domain optical coherence tomography of the human retina and optic nerve. Optics Express. 2004;**12**(3): 367-376. DOI: 10.1364/OPEX.12.000367

[9] Mogensen M, Thrane L, Jørgensen TM, Andersen PE, Jemec GBE. OCT imaging of skin cancer and other dermatological diseases. Journal of Biophotonics. 2009;**2**(6–7): 442-451. DOI: 10.1002/jbio.200910020

[10] Mogensen M, Nürnberg BM, Forman JL, Thomsen JB, Thrane L, Jemec GBE. In vivo thickness measurement of basal cell carcinoma and actinic keratosis with optical coherence tomography and 20-MHz ultrasound. The British Journal of Dermatology. 2009;**160**(5):1026-1033. DOI: 10.1111/j.1365-2133.2008.09003.x

[11] Shen K, Lu H, Baig S, Wang MR. Improving lateral resolution and image quality of optical coherence tomography by the multi-frame superresolution technique for 3D tissue imaging. Biomedical Optics Express. 2017;**8**(11): 4887-4918. DOI: 10.1364/BOE.8.004887

[12] Shen K, Lu H, Wang MR. Improving lateral resolution of optical coherence tomography for imaging of skins. Proceedings of SPIE. 2016;**9713**:97130N. DOI: 10.1117/12.2213143

[13] Yazdanfar S, Rollins AM, Izatt JA. In vivo imaging of human retinal flow dynamics by color Doppler optical coherence tomography. JAMA Ophthalmology. 2003;**121**(2):235-239. DOI: 10.1001/archopht.121.2.235

[14] Chu CR, Izzo NJ, Irrgang JJ, Ferretti M, Studer RK. Clinical diagnosis of potentially treatable early articular cartilage degeneration using optical coherence tomography. Journal of Biomedical Optics. 2007;**12**(5):051703. DOI: 10.1117/1.2789674

[15] Stifter D. Beyond biomedicine: A review of alternative applications and developments for optical coherence tomography. Applied Physics B. 2007; **88**(3):337-357. DOI: 10.1007/s00340-007-2743-2

[16] Demos SG, Staggs M, Minoshima K, Fujimoto J. Characterization of laser induced damage sites in optical components. Optics Express. 2002; **10**(25):1444-1450. DOI: 10.1364/OE.10.001444

[17] Heise B, Schausberger SE, Häuser S, Plank B, Salaberger D, Leiss-Holzinger E, et al. Full-field optical coherence microscopy with a sub-nanosecond supercontinuum light source for material research. Optical Fiber Technology. 2012;**18**(5):403-410. DOI: 10.1016/j.yofte.2012.07.011

[18] Stifter D, Leiss-Holzinger E, Major Z, Baumann B, Pircher M, Götzinger E, et al. Dynamic optical studies in materials testing with spectral-domain polarization-sensitive optical coherence tomography. Optics Express. 2010;**18**(25):25712-25725. DOI: 10.1364/OE.18.025712

[19] Shen K, Lu H, Wang JH, Wang MR. Improved resolution of optical coherence tomography for imaging of microstructures. Proceedings of SPIE. 2015;**9334**:93340X. DOI: 10.1117/12.2080415

[20] Shen K, Jiang G, Mao W, Baig S, Wang MR. Fabrication of two kinds of eight-fold photonic quasi-crystals assisted by a specially designed prism. Applied Optics. 2013;**52**(26):6474-6480. DOI: 10.1364/AO.52.006474

[21] Jiang G, Shen K, Wang MR. Fabrication of 3D Micro- and Nano-Structures by Prism-Assisted UV and Holographic Lithography. Rijeka, Croatia: Intech; 2013. pp. 227-252. DOI: 10.5772/56417

[22] Shen K, Jiang G, Wang MR. Eight-fold photonic quasi-crystal fabricated by prism assisted holographic lithography. Proceedings of SPIE. 2013;**8632**:86320G. DOI: 10.1117/12.2004355

[23] Mao W, Shen K, Wang MR. Holographic eight-fold photonic quasi-crystal. Proceedings of SPIE. 2012;**8269**: 82691R. DOI: 10.1117/12.908654

[24] Alarousu E, Krehut L, Prykäri T, Myllylä R. Study on the use of optical coherence tomography in measurements of paper properties. Measurement Science and Technology. 2005;**16**(5):1131-1137. DOI: 10.1088/0957-0233/16/5/012

[25] Targowski P, Ra M, Wojtkowski M. Optical coherence tomography for artwork diagnostics. Laser Chemistry. 2006;**2006**:35373. DOI: 10.1155/2006/35373

[26] Koller DM, Hannesschläger G, Leitner M, Khinast JG. Non-destructive analysis of tablet coatings with optical coherence tomography. European Journal of Pharmaceutical Sciences. 2011;**44**(1):142-148. DOI: 10.1016/j.ejps.2011.06.017

[27] Wang JH, Wang MR. Handheld non-contact evaluation of fastener flushness and countersink surface profiles using optical coherence tomography. Optics Communications. 2016;**371**:206-212. DOI: 10.1016/j.optcom.2016.03.069

[28] Liu M, Buma T. Biometric mapping of fingertip eccrine glands with optical coherence tomography. IEEE Photonics Technology Letters. 2010;**22**(22): 1677-1679. DOI: 10.1109/LPT.2010.2079926

[29] Darlow LN, Connan J, Akhoury SS. Internal fingerprint zone detection in optical coherence tomography fingertip scans. Journal of Electronic Imaging.

2015;**24**(2):023027. DOI: 10.1117/1.JEI.24.2.023027

[30] Liu X, Zaki F, Wang Y, Huang Q, Mei X, Wang J. Secure fingerprint identification based on structural and microangiographic optical coherence tomography. Applied Optics. 2017; **56**(8):2255-2259. DOI: 10.1364/AO.56.002255

[31] Shen K, Baig S, Palacios D, Wang MR. 3D high-resolution subsurface fingerprint imaging using superresolution optical coherence tomography. Proceedings of SPIE. 2019; **10867**:108673M. DOI: 10.1117/12.2511812

[32] Wang Y, Zhao Y, Nelson JS, Chen Z, Windeler RS. Ultrahigh-resolution optical coherence tomography by broadband continuum generation from a photonic crystal fiber. Optics Letters. 2003;**28**(3):182-184. DOI: 10.1364/OL.28.000182

[33] Shidlovski VR. Superluminescent diode light sources for OCT. In: Drexler W, Fujimoto JG, editors. Optical Coherence Tomography: Technology and Applications. Cham: Springer International Publishing; 2015. pp. 505-526. DOI: 10.1007/978-3-319-06419-2_18

[34] Izatt JA, Choma MA. Theory of optical coherence tomography. In: Drexler W, Fujimoto JG, editors. Optical Coherence Tomography: Technology and Applications. Berlin/Heidelberg: Springer; 2008. pp. 47-72. DOI: 10.1007/978-3-540-77550-8_2

[35] Fercher AF, Drexler W, Hitzenberger CK, Lasser T. Optical coherence tomography—Principles and applications. Reports on Progress in Physics. 2003;**66**(2):239-303. DOI: 10.1088/0034-4885/66/2/204

[36] Szkulmowski M, Gorczynska I, Szlag D, Sylwestrzak M, Kowalczyk A, Wojtkowski M. Efficient reduction of speckle noise in optical coherence tomography. Optics Express. 2012; **20**(2):1337-1359. DOI: 10.1364/OE.20.001337

[37] Shen K, Baig S, Jiang G, Wang MR. Wide field-of-view high lateral resolution optical coherence tomography with 3D image stitching and superresolution. Proceedings of SPIE. 2018;**10483**:104832R. DOI: 10.1117/12.2287480

[38] Shen K. High Resolution Wide Field-of-View Optical Coherence Tomography for 3D Imaging. Coral Gables: University of Miami; 2018. Available from: https://scholarlyrepository.miami.edu/oa_dissertations/2076

[39] Zawadzki RJ, Jones SM, Olivier SS, Zhao M, Bower BA, Izatt JA, et al. Adaptive-optics optical coherence tomography for high-resolution and high-speed 3D retinal *in vivo* imaging. Optics Express. 2005;**13**(21):8532-8546. DOI: 10.1364/OPEX.13.008532

[40] Klein T, Wieser W, Eigenwillig CM, Biedermann BR, Huber R. Megahertz OCT for ultrawide-field retinal imaging with a 1050 nm Fourier domain mode-locked laser. Optics Express. 2011;**19**(4):3044-3062. DOI: 10.1364/OE.19.003044

[41] Liu Z, Kocaoglu OP, Miller DT. In-the-plane design of an off-axis ophthalmic adaptive optics system using toroidal mirrors. Biomedical Optics Express. 2013;**4**(12):3007-3030. DOI: 10.1364/BOE.4.003007

[42] Wang B, Lu R, Zhang Q, Yao X. Breaking diffraction limit of lateral resolution in optical coherence tomography. Quantitative Imaging in Medicine and Surgery. 2013;**3**(5):243-248. DOI: 10.3978/j.issn.2223-4292.2013.10.03

[43] Robinson MD, Chiu SJ, Lo J, Toth C, Izatt J, Farsiu S. New applications of

super-resolution in medical imaging. Super-Resolution Imaging. 2010: 384-412

[44] Agrawal A, Connors M, Beylin A, Liang C-P, Barton D, Chen Y, et al. Characterizing the point spread function of retinal OCT devices with a model eye-based phantom. Biomedical Optics Express. 2012;**3**(5):1116-1126. DOI: 10.1364/BOE.3.001116

[45] Ralston TS, Marks DL, Kamalabadi F, Boppart SA. Deconvolution methods for mitigation of transverse blurring in optical coherence tomography. IEEE Transactions on Image Processing. 2005;**14**(9):1254-1264. DOI: 10.1109/TIP.2005.852469

[46] Liu Y, Liang Y, Mu G, Zhu X. Deconvolution methods for image deblurring in optical coherence tomography. Journal of the Optical Society of America A. 2009;**26**(1):72-77. DOI: 10.1364/JOSAA.26.000072

[47] Hojjatoleslami SA, Avanaki MRN, Podoleanu AG. Image quality improvement in optical coherence tomography using Lucy-Richardson deconvolution algorithm. Applied Optics. 2013;**52**(23):5663-5670. DOI: 10.1364/AO.52.005663

[48] Sander B, Larsen M, Thrane L, Hougaard JL, Jørgensen TM. Enhanced optical coherence tomography imaging by multiple scan averaging. British Journal of Ophthalmology. 2005;**89**(2): 207-212. DOI: 10.1136/bjo.2004.045989

[49] Rogowska J, Brezinski ME. Image processing techniques for noise removal, enhancement and segmentation of cartilage OCT images. Physics in Medicine and Biology. 2002; **47**(4):641-655. DOI: 10.1088/0031-9155/47/4/307

[50] Izatt JA, Choma MA, Dhalla A-H. Theory of optical coherence tomography. In: Drexler W, Fujimoto JG, editors. Optical Coherence Tomography: Technology and Applications. Cham: Springer International Publishing; 2015. pp. 65-94. DOI: 10.1007/978-3-319-06419-2_3

[51] Hecht E. Optics. Boston: Addison-Wesley Publishing Company Incorporated; 2016

[52] Yang J, Huang T. Image super-resolution: Historical overview and future challenges. Super-Resolution Imaging. 2010:1-34

[53] Farsiu S, Robinson MD, Elad M, Milanfar P. Fast and robust multiframe super resolution. IEEE Transactions on Image Processing. 2004;**13**(10): 1327-1344. DOI: 10.1109/TIP.2004.834669

[54] Park SC, Park MK, Kang MG. Super-resolution image reconstruction: A technical overview. Signal Processing Magazine, IEEE. 2003;**20**(3):21-36. DOI: 10.1109/MSP.2003.1203207

[55] Rafael Gonzalez C, Woods R. Digital Image Processing. New Jersey: Pearson Education; 2002

[56] Knutsson H, Westin C-F, editors. Normalized and differential convolution. In: 1993 IEEE Computer Society Conference on Computer Vision and Pattern Recognition (CVPR'93). IEEE; 1993. DOI: 10.1109/CVPR.1993.341081

[57] Pham TQ, Van Vliet LJ, Schutte K. Robust fusion of irregularly sampled data using adaptive normalized convolution. EURASIP Journal on Advances in Signal Processing. 2006; **2006**(1):1-12. DOI: 10.1155/ASP/2006/83268

[58] Farnebäck G. Polynomial Expansion for Orientation and Motion Estimation. Linköping: Linköping University Electronic Press; 2002

[59] Richardson WH. Bayesian-based iterative method of image restoration. Journal of the Optical Society of America. 1972;**62**(1):55-59. DOI: 10.1364/JOSA.62.000055

[60] Lucy LB. An iterative technique for the rectification of observed distributions. The Astronomical Journal. 1974;**79**:745. DOI: 10.1086/111605

[61] Fergus R, Singh B, Hertzmann A, Roweis ST, Freeman WT. Removing Camera Shake from a Single Photograph. New York: ACM Transactions on Graphics (TOG)—ACM; 2006. DOI: 10.1145/1141911.1141956

[62] Krishnan D, Tay T, Fergus R, editors. Blind deconvolution using a normalized sparsity measure. In: 2011 IEEE Conference on Computer Vision and Pattern Recognition (CVPR). IEEE; 2011. DOI: 10.1109/CVPR.2011.5995521

[63] Hecht E. Optics. 4th ed. Reading: Addison Wesley Longman Inc; 1998

[64] Wells WM, Viola P, Atsumi H, Nakajima S, Kikinis R. Multi-modal volume registration by maximization of mutual information. Medical Image Analysis. 1996;**1**(1):35-51. DOI: 10.1016/S1361-8415(01)80004-9

[65] Duda RO, Hart PE. Pattern Classification and Scene Analysis. New York: Wiley; 1973

[66] Papoulis A. Random Variables, and Stochastic Processes. New York: McGraw-Hill; 1990

[67] Jian Z, Yu L, Rao B, Tromberg BJ, Chen Z. Three-dimensional speckle suppression in optical coherence tomography based on the curvelet transform. Optics Express. 2010;**18**(2):1024-1032. DOI: 10.1364/OE.18.001024

[68] Kuroda T. Essential Principles of Image Sensors. Boca Raton: CRC Press; 2014

[69] Kraus MF, Hornegger J. OCT motion correction. In: Optical Coherence Tomography: Technology and Applications. Cham: Springer International Publishing; 2015. pp. 459-476. DOI: 10.1007/978-3-319-06419-2_16

[70] Levin A, Weiss Y, Durand F, Freeman WT. Understanding blind deconvolution algorithms. IEEE Transactions on Pattern Analysis and Machine Intelligence. 2011;**33**(12):2354-2367. DOI: 10.1109/TPAMI.2011.148

[71] Jiang G, Baig SA, Lu H, Shen K, Wang MR. Planar concave grating wavelength demultiplexers with flattened spectral response based on SU-8 polymer waveguides. Optical Engineering. 2016;**55**(3):1-6. DOI: 10.1117/1.OE.55.3.037104

[72] Jiang G, Baig S, Lu H, Shen K, Wang MR. Planar concave grating with flattened spectral response for wavelength demultiplexing optical interconnection. Proceedings of SPIE. 2015;**9368**:93680R. DOI: 10.1117/12.2080380

[73] Shen K, Baig S, Jiang G, Paik Y-H, Kim SJ, Wang MR. Improved light emitting UV curable PbS quantum dots-polymer composite optical waveguides. Optics Communications. 2017;**402** (Supplement C):606-611. DOI: 10.1016/j.optcom.2017.06.083

[74] Shen K, Baig S, Jiang G, Paik Y-H, Kim SJ, Wang MR. Broadband infrared light emitting waveguides based on UV curable PbS quantum dot composites. Proceedings of SPIE. 2018;**10538**:105380L. DOI: 10.1117/12.2287464

[75] Bay H, Ess A, Tuytelaars T, Van Gool L. Speeded-up robust features (SURF). Computer Vision and Image Understanding. 2008;**110**(3):346-359. DOI: 10.1016/j.cviu.2007.09.014

[76] Schouten B, Jacobs B. Biometrics and their use in e-passports. Image and Vision Computing. 2009;**27**(3): 305-312. DOI: 10.1016/j.imavis.2008.05.008

[77] Zam A, Dsouza R, Subhash HM, O'Connell M-L, Enfield J, Larin K, et al. Feasibility of correlation mapping optical coherence tomography (cmOCT) for anti-spoof sub-surface fingerprinting. Journal of Biophotonics. 2013;**6**(9):663-667. DOI:10.1002/jbio.201200231

[78] Daugman J. How iris recognition works. IEEE Transactions on Circuits and Systems for Video Technology. 2004;**14**(1):21-30. DOI: 10.1109/TCSVT.2003.818350

[79] Chang KI, Bowyer KW, Flynn PJ. An evaluation of multimodal 2D+3D face biometrics. IEEE Transactions on Pattern Analysis and Machine Intelligence. 2005;**27**(4):619-624. DOI: 10.1109/TPAMI.2005.70

Chapter 3

OCT in Applications That Involve the Measurement of Large Dimensions

Nélida A. Russo, Eneas N. Morel, Jorge R. Torga and Ricardo Duchowicz

Abstract

The application of optical coherence tomography (OCT) technique is not very common when measuring large dimensions is required. This type of measurements can be critical to achieve satisfactory results in the manufacturing process of precision parts. Components and structures ranging from submillimeter to several centimeters size can be found in many fields including automotive, aerospace, semiconductor, and data storage industries to name a few. In this chapter, an interferometric system based on the swept source optical coherence tomography (SS-OCT) technique, which has a wide measurement range and good axial resolution, is presented and its constituent parts are analyzed. The scheme includes a self-calibration stage based on fiber Bragg gratings (FBGs) that allows monitoring the spectral position of the light source in each scan, having the advantage of being a passive system that requires no additional electronic devices. Several applications of the system are described, including measurement of distances up to 17 cm, characterization of multilayer transparent and semitransparent structures, simultaneous determination of thickness of the wall, internal and external diameter of glass ampoules or similar containers, thickness measurements in opaque samples or where the refractive index is unknown, etc.

Keywords: interferometry, optical coherence tomography, optical metrology, nondestructive testing, large axial range

1. Introduction

Optical coherence tomography (OCT) is a noninvasive, interferometric technique that provides real-time 3-D images with micrometric resolution and depth of penetration that can range from some millimeters to a few centimeters, depending on the technique employed and the material under study. OCT images provide structural information of a sample, based on backscattered light from different layers of material within it. This technique is considered the optical analogue to ultrasound; however, it achieves a higher resolution using near-infrared wavelengths, at the cost of decreasing depth of penetration.

OCT was developed in the early 1990s for the noninvasive imaging of biological tissue [1, 2]. The first application was in ophthalmology [3] where it has had great development [4–6], the same as in cardiology [7, 8], where commercial equipment has already been developed. In biomedicine, the use of relatively long-wavelength light allows it to penetrate into the scattering medium.

In the last years, low-coherence interferometric techniques such as OCT have been proposed as a powerful tool for industrial nondestructive testing (NDT), even leading to the development of specific equipment. Among the industrial metrological applications, the following can be mentioned: the measurement of very thin thicknesses in semiconductor wafers [9, 10], the characterization of surfaces [11, 12], the control of thickness in the coating of pills in the pharmaceutical industry [13], and others. OCT systems based on fiber optics are particularly suitable for use in industrial [14], hostile (electromagnetic interference, radioactive, cryogenic, or very high temperatures) [15], or difficult-to-access environments [16]. Moreover, optical fiber-based OCT devices can take the advantage of beam stability and lower price of IR components used in optical communications. However, applications such as 3-D inspection in large parts (greater than a centimeter) constitute a little explored area, since its implementation still has some technological limitations [17]. When using time-domain OCT, the interference signal is generated from the displacement of a mirror in the reference arm of the interferometer to equalize the sample distances. For measuring great dimensions, it implies the use of very expensive, extremely accurate, and with high resolution moving mechanical systems. On the other hand, spectral or Fourier-domain OCT allows the design of an interferometric system more robust, compact, and faster [18], thereby increasing the mechanical stability and reducing the sensitivity to vibrations. However, since it uses a spectrometer to detect the interference signal, its spectral resolution limits the measuring range below 1 cm. The SS-OCT technique offers an interesting alternative provided by the use of tunable light sources [19–21] such as fiber-optic lasers with variable spacing Fabry-Perot filters or electro-optical modulators [22] or tunable semiconductor lasers by intracavity refractive index change or by the use of microelectromechanical system (MEMS) cavities [23]. In general, in these systems, it is possible to achieve a laser linewidth narrower than the resolutions typically obtained by a spectrometer. Consequently, the depth range, in this case given by the instantaneous linewidth of the laser (coherence length), can reach values in the order of 1 cm or more and greatly expand the potential industrial applications [24–28]. In this case, the detection system is a photodetector in conjunction with a digitizer or oscilloscope, simplifying the system. In general, scanning sources have the inherent problem of not being able to provide accurate information about the correspondence between the wavelength, the signal voltage applied to the tuning device, and the temporary location within each sweep [29]. Some solutions have been proposed [30, 31], but in general they are complex and expensive.

In this context, this chapter reports and discusses a simple and self-calibrated fiber-optic interferometric system based on the swept source optical coherence tomography (SS-OCT) technique, especially suitable when the measurement of large dimensions (several centimeters) is necessary as those ones involved in industrial metrological applications [32–37]. Its constituent parts are analyzed, and several applications are shown, including the measurement of several centimeter distances, characterization of multilayer structures, simultaneous determination of thickness of the wall, internal and external diameter of glass containers, thickness measurements in opaque samples (which allows the use of the system to perform profilometry of mechanical parts) or where the refractive index is unknown, etc. On the other hand, the use of a set of fiber Bragg gratings to relate the emission wavelength of the tunable laser source and the temporary position in each sweep is discussed. The factors that determine the resolution and the maximum range of distances to be measured are presented. It is shown that with this system it is possible to determine distances of up to 17 cm with a spatial resolution in the order of 21 microns, which constitutes a very encouraging scheme for the dimensional inspection applications mentioned before.

2. Description of the measurement scheme

The measuring scheme used in this work is essentially composed of the stages or subsystems indicated in **Figure 1**. The light source, the self-calibration system, and the interferometric system used for the different applications were implemented using monomode optical fiber operating in the 1550 nm spectral region. In the following subsections, each of them will be discussed in detail.

Basically, an erbium-doped fiber laser tunable in the 1550 nm spectral region was used as the light source. It has sufficient coherence length to allow the formation of great depth images. A passive self-calibration stage based on fiber Bragg gratings allows counteracting the possible variation in the scanning speed of the laser caused by the tuner module, in addition to linking the emission wavelength of the laser and the temporal position in each sweep. The interferometric system consists of a Michelson-type arrangement for the measurement of distances and transparent object thickness or a ring interferometer with a Sagnac-Michelson configuration for the measurement of opaque samples (or with unknown refractive index). The temporal distribution of the interference signals was obtained using an InGaAs photodetector with 2 GHz bandwidth connected to a digital oscilloscope, although it can be replaced by any 200 MHz bandwidth DAC system and two acquisition channels. As will be explained in detail later, in order to carry out the processing of the detected signal while the light source sweeps in wavelength, the Fourier transform was applied, and the position of the interfering peaks allowed obtaining the desired distance or thickness.

2.1 Laser system

The light source was a continuous wave emission erbium-doped fiber laser, tunable in a spectral range from about 1520 to 1570 nm using a variable spaced Fabry-Perot filter (Micron Optics, FFP-TF2) that has a free spectral range (FSR) of 60 nm and a bandwidth of 60 pm. Sweeping is achieved by applying a periodical signal like a triangular voltage waveform to the filter cavity PZT actuator [38]. By varying the parameters (amplitude and offset) of this signal, it is possible to modify the tuning range of the fiber laser and consequently the value of the maximum measurable distance. The sweep frequency could be varied up to 100 Hz. The erbium-doped fiber was pumped by a semiconductor laser diode emitting at 980 nm. The length of the doped fiber was selected taking into account a trade-off between spectral emission flatness and output power. The fiber laser output beam was extracted from the ring cavity through the 10% port of a 10/90 optical coupler. **Figure 2** shows the laser emission as it sweeps the tuning range, recorded by an optical spectrum analyzer (OSA) (Yokogawa, model AQ6370B). Since the acquisition speed of this instrument is much slower than the speed at which the F-P filter can be moved, the OSA records several sweeps of the laser finding it in different

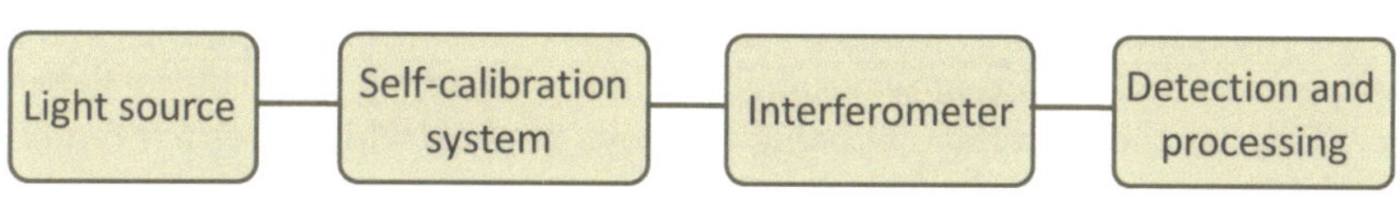

Figure 1.
Block diagram of the measuring system.

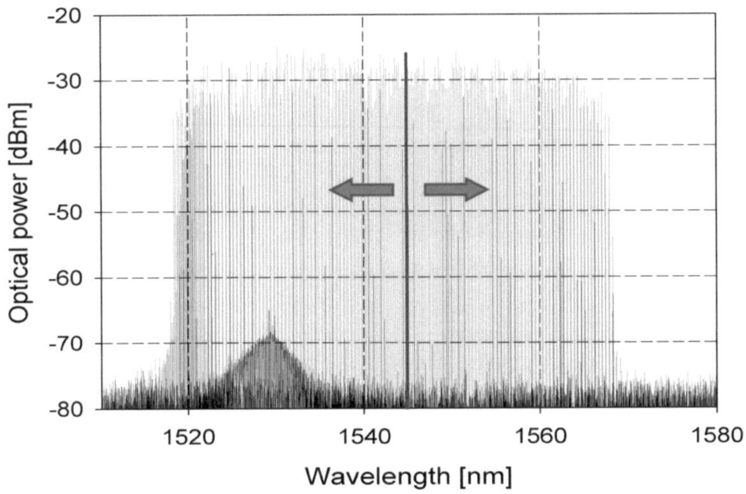

Figure 2.
Scanning spectrum of the laser source.

spectral positions. One of them is highlighted and the arrows represent the sweeping of the emission. As can be seen, despite the relatively low output power, the fiber laser has a very good S/N ratio. When the laser operates at any fixed wavelength within the tuning range, the emission spectrum has a typical width of 20 pm at 3 dB.

By considering a rectangular spectral profile when the laser source is tuned in the abovementioned range, the maximum theoretical axial resolution (or depth resolution) of the measurement system can reach 21 μm, which is derived from the following expression (Eq. (1)):

$$Res = \frac{2\ln(2)}{\pi}\frac{\lambda_0^2}{\Delta\lambda} \tag{1}$$

where λ_0 is the central wavelength of the source spectrum and $\Delta\lambda$ is the spectral width of the broadband light source or, in our case, we should consider the tuning range. However, depending on the requirements, it is possible to modify (decrease) the swept range by means of the FP filter controller, which obviously worsens the axial resolution.

2.2 Calibration of the laser emission

In applications of low-coherence interferometry by swept source, the linear sampling process of spatial frequency (k-space) is critical. In the system used in this work, the tuning of the laser is done by controlling the cavity of a Fabry-Perot filter, by applying a triangular signal to a piezoelectric actuator. This process is affected by different factors such as hysteresis of the piezoelectric actuator (PZT) and errors in the repetition control of the scan cycle, which may make the spectral position of the laser emission unknown at each instant of time. Thus, the use of a self-calibration system is proposed [32, 33], which consists of a set of fiber Bragg gratings (FBGs) [39–41] centered on different wavelengths within the tuning range of the light source. This has the advantage of being a passive system, which does not require additional electronic devices. This system allows relating the emission wavelength of the source and the temporal position in each sweep. Since the spectral locations of the FBGs are known, from the measurement of their temporal positions, it is possible to calibrate the OCT measurement system, resizing for each sweep of the source the time axis provided by the oscilloscope used to acquire the signal. The objective is to obtain a sequence of values k_{FBGi} (wave number corresponding to the ith filter) and the associated times (t_{FBGi}) within the whole scanning range. In this way, two vectors k_{FBGi} and t_{FBGi} can be generated (Eqs. (2) and (3)):

$$k_{FBG} = [k_{FBG_1}, k_{FBG_2}, k_{FBG_3}, \ldots, k_{FBGn}] \tag{2}$$

$$t_{FBG} = [t_{FBG_1}, t_{FBG_2}, t_{FBG_3}, \ldots, t_{FBGn}] \tag{3}$$

With these values of t_{FBG} and k_{FBG}, a curve is obtained that represents the nonlinearity in the movement of PZT. From an interpolation, we obtain the function $k(t)$ that relates the sampling times to a uniformly sampled k-space, and from it we obtain the sampling frequency to be used in the calculation of the Fourier transform (Eq.(4)):

$$f_s = \frac{1}{(k_{n-1} - k_n)} \tag{4}$$

The error of the frequency f_s is obtained from the propagation of errors in the following way:

$$\Delta f_s = \left| \frac{\partial f_s}{\partial \Delta k} \right| \Delta k \tag{5}$$

In principle, the number of FBGs to use is a trade-off between obtaining the best calibration curve to minimize adjustment errors and generating the least possible deterioration in the detected signal due to insertion losses and noise caused by backward reflections.

As it is known, FBGs are sensitive to thermal changes, so variations in the temperature of the environment induce changes in their spectral positions. However, since all the gratings that integrate our calibration system are recorded in the same type of optical fiber and all of them experience the same temperature change, this will not generate measurement errors of the OCT system. This is because to determine the value of the sampling frequency, it is only relevant that the value Δk between FBGs remains constant, even though their individual positions (k_i) may change. The change in the absolute spectral positions of the Bragg gratings due to thermal variations only generates a shift in the k-space and does not affect the determination of the sampling frequency. On the other hand, gratings do not need to be equally spaced within the range of laser tuning. The central wavelengths of the FBGs employed in our work were 1527.84, 1535.73, 1541.82, 1547.65, 1553.62, and 1558.43 nm, and their spectral widths were in the order of 0.1 nm.

2.3 Fiber-optic interferometric system

2.3.1 Michelson interferometer

The configuration used for the measurement of distances or thicknesses of transparent samples consists of a Milchelson-type interferometric system that is shown schematically in **Figure 3**.

If the light source has a spectral profile $S(k)$ and considering the interaction between the reference arm (RA) and the first interface (L_1) of the sample arm (SA), the interference signal is determined by

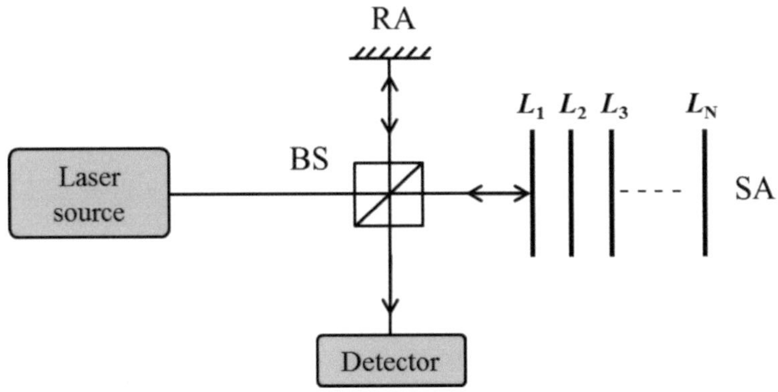

Figure 3.
Basic configuration of the Michelson-type interferometer. BS, beam splitter; RA, reference arm; SA, sample arm; $L_1 \ldots L_N$, *different interfaces of the sample.*

$$I(k) = S(k)\left[I_r + I_{L_1} + 2\sqrt{I_r I_{L_1}}\cos\left(k(d_r - d_{L_1})\right)\right] \tag{6}$$

where $d_r - d_{L_1} = n\,\Delta d$ is the optical path difference (OPD) that light travels between the RA and the L_1 interface of the sample and n is the group refractive index of the medium. It is evident that the cosine frequency indicates the value of $n\,\Delta d$. This type of signal is obtained when the system is used to measure the distance to an opaque surface.

If a sample with multiple interfaces (L_1 to L_N) is considered, the multiple reflections that appear will generate different terms or interference components, and the total intensity can be described by the following equation:

$$\begin{aligned} I(k) = S(k)\Bigg[I_r &+ \sum_{n=1}^{N} I_{L_n} + \sum_{n=1}^{N} 2\sqrt{I_r\, I_{L_n}}\cos\left(k\,(d_r - d_{L_n})\right) \\ &+ \sum_{n\neq m=1}^{N} 2\sqrt{I_{L_n}\, I_{L_m}}\cos\left(k\,(d_{L_m} - d_{L_n})\right)\Bigg] \end{aligned} \tag{7}$$

The first term of Eq. (7) is the intensity reflected in the mirror of the reference arm, and the second is the sum of the intensities reflected in the different interfaces of the sample. These two terms are often referred to as "constant" or "DC" components. The first one generates the largest contribution to the detected intensity if the reflectivity of the reference dominates over the reflectivity of the different layers of the sample. The second sum is called "cross-correlation" for each reflector of the sample, and it depends on both the wave number of the light source and the OPD between the reference arm and the reflector considered in the sample arm. These terms are usually smaller than the DC part. However, the dependence of the square root represents an important logarithmic gain factor over the direct detection of the reflections of the sample. The third sum is called "autocorrelation" and represents the interference that occurs between the different interfaces present in the sample. Since the autocorrelation terms depend linearly on the reflectivity of each layer of the sample which cannot be controlled, a way to manage their intensities is the appropriate selection of the reference reflectivity. In the last two sums, the direct interactions of each interface with the reference arm (d_r-d_{Ln}) and the additional interactions between the different interfaces of the medium to be analyzed (d_{Lm}-d_{Ln}) are evident. It should be clarified that in this case the different OPDs can be linked to different refractive index depending on the medium through which the light propagates. Then, it is necessary to consider the refractive index that corresponds to each layer.

As regards the fiber-optic implementation of the aforementioned interferometric system, two variants were used, which can be seen in **Figure 4**. In the case of distance measurement, the light beam is divided by a fiber-optic coupler (FC 50/50). One of the beams is directed to the reference surface (a mirror) and the other to the sample. After being reflected in each of these surfaces, both beams are directed to the detection system, where their superposition generates the interference signal. From its processing, the optical path difference (OPD) between both arms is obtained and, consequently, the unknown distance. When it is desired to measure transparent or semitransparent samples, light is reflected from subsurface structures within it (e.g., from both sides of the sample if its thickness is being measured or from the different interfaces if it is a multilayer sample), so it is possible to perform a tomography measurement.

2.3.2 Sagnac-Michelson interferometer

For the measurement of opaque samples, it is necessary to modify the interferometric system and use a configuration based on a combination of the Michelson interferometer with the Sagnac interferometer. This configuration allows the determination of thicknesses of opaque samples or whose refractive index is unknown, as well as the surface characterization of nontransparent objects (profilometry of mechanical parts).

Figure 5 shows the configuration of the Sagnac-Michelson interferometer in optical fiber, where E1 is the mirror of the reference arm, while BS1 and BS2 are beam splitters implemented with fused fiber-optic couplers with a coupling ratio 50:50. The beams that illuminate both the mirror in the reference arm and both sides of the sample are collimated by means of two single-mode GRIN fiber-optic collimators (Col.). The ring that forms at the output of BS2 with the beams that illuminate the sample forms the Sagnac configuration.

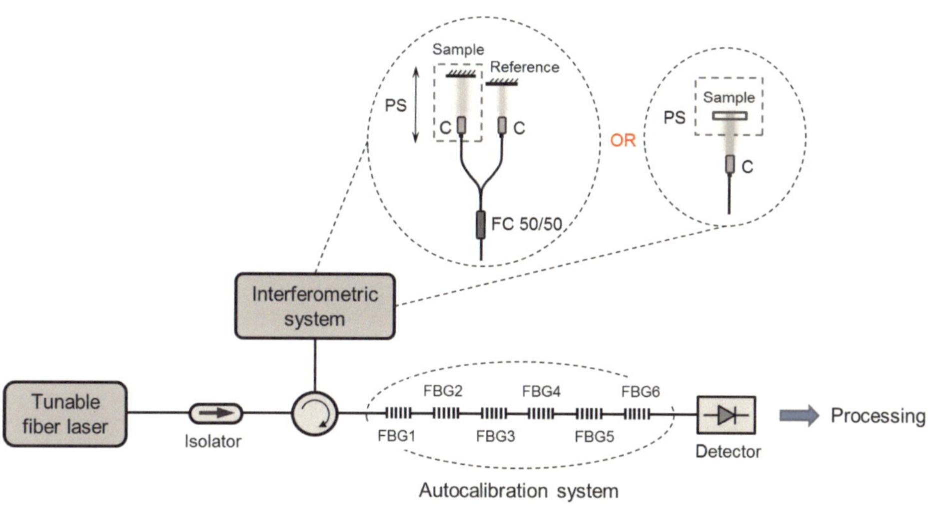

Figure 4.
Detailed setup employed to measure distances and thicknesses of multilayer transparent samples. PS, positioning system; C, collimator; FC, fiber coupler.

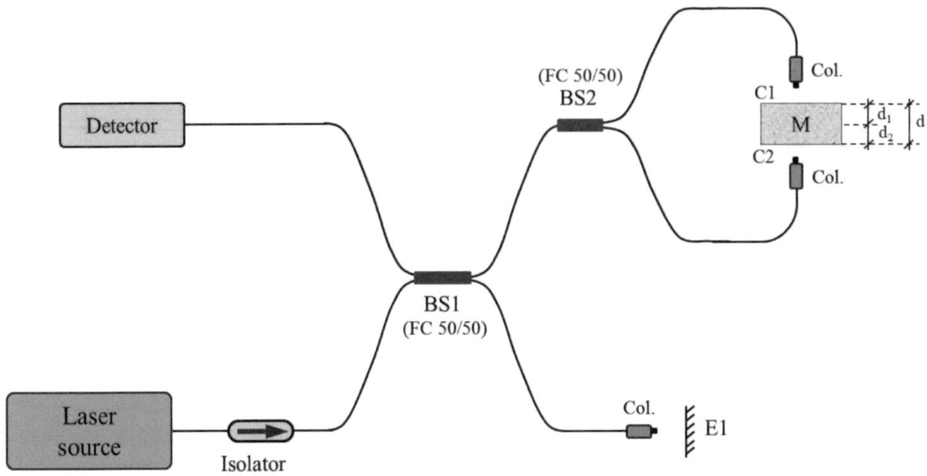

Figure 5.
Sagnac-Michelson interferometer implemented in optical fiber.

The path performed by the light beam can be described as follows. After the first splitter (BS1), one beam travels toward the reference (E1) and the other toward the second splitter (BS2), where it is separated again into two beams, each of which illuminates opposite faces of the sample. Upon reaching the sample (M), the beams are reflected and perform the reverse way. The reflections on the three surfaces, the mirror E1 and the two faces of the sample C1 and C2, are superimposed on the detector and produce the interference signal containing the information of interest (the thickness of the sample in the direction of illumination).

The optical path differences between the distances traveled by the different light beams generate the modulations in the interference signal, which are analyzed in the Fourier space. From this analysis, the OPDs can be determined in each case, obtaining the position of the maximum of the "interference peak" in the Fourier transform.

We will call P_r as the position (in the OPD axis) of the peak corresponding to the term (or component) of the interference signal generated by the reflection at E1 (reference beam) and the beam traveling through the Sagnac interferometer when there is no sample (and with the corresponding collimators aligned); P_1 as the peak that is produced by the interference between the reference and the reflection on the face C1 of the sample; P_2 as the interference peak that occurs between the reference and the reflection on the other side of the sample (C2). The distances L_i represent the paths traveled by the light in the different sectors of the interferometer (see **Figure 5**), that is, L_1 is the optical path between BS2 and the face C1 of the sample, L_2 is the optical path between BS2 and the face C2 of the sample, L_3 is the optical path between BS1 and BS2, LR is the length of the reference arm (between BS1 and E1), and d is the thickness of the sample. In addition, d_1 and d_2 are the segments of the thickness of the sample (d) on each side of the axis of symmetry of the Sagnac interferometer. Therefore,

$$d = d_1 + d_2 \tag{8}$$

$$P_r = 2LR - (2L_3 + L_1 + L_2 + d) \tag{9}$$

$$P_1 = 2LR - 2(L_3 + L_1) \tag{10}$$

$$P_2 = 2LR - 2(L_3 + L_2) \tag{11}$$

By operating with the above equations, it is possible to determine the thickness of the sample from

$$(P_1 + P_2) - 2P_r = 2d \tag{12}$$

From this last expression, it is possible to determine the thickness of the sample under study. It is interesting to note that it was not necessary to make considerations regarding the lengths that the interferometer branches should have or the position of the sample so that Eq. (12) is general and can be used in any condition.

2.4 Detection and processing

The temporal distribution of the optical intensity corresponding to the interference signal was obtained using an InGaAs photodetector connected to a digital oscilloscope. **Figure 6** shows a typical record of (a) a triangular voltage signal used to tune the fiber laser, (b) an interferometric signal that contains the dimensional information of the sample, and (c) the calibration signal with FBG attenuation peaks used to relate the emission wavelength of the source and the temporal position in each sweep.

Basically, the Fourier transform was then applied to the detected signal, and the position of the interference peak allowed obtaining the desired distance or thickness. As was mentioned before, the interference signal is affected by uncertainties that arise from the nonlinear movement of the PZT used to sweep the laser. To overcome this problem, fiber Bragg gratings of known spectral positions were used, which allowed transforming the temporal axis into an axis in wave number (k). To apply the fast Fourier transform (FFT) that reconstructs the axial scan as a function of depth, the spectrum must be sampled uniformly in the k-space. Therefore, a linearization of the k-vector was performed, and the interference signal was interpolated to obtain a sampled signal at equally spaced intervals. The resampled interference signal can be expressed as indicated in Eq. (13):

$$I(k) = S(k)\left[I_1 + I_2 + 2\sqrt{I_1 I_2}\cos\left(k(nd)\right)\right] \tag{13}$$

where k is the wave number, $S(k)$ is the spectral profile of the light source, I_1 and I_2 are the amplitudes of the intensities reflected in both arms of the interferometer (to simplify the notation, their dependence with k was not indicated), and $n\,d$ is the optical path difference (with n being the group refractive index of the medium in which light is propagated and d being the difference in length between both interferometer arms).

To illustrate the procedure used, the graphs obtained by measuring a thickness of d = 500 microns and n = 1 are shown. In that case, the interference signal in the k-space has the shape shown in **Figure 7**.

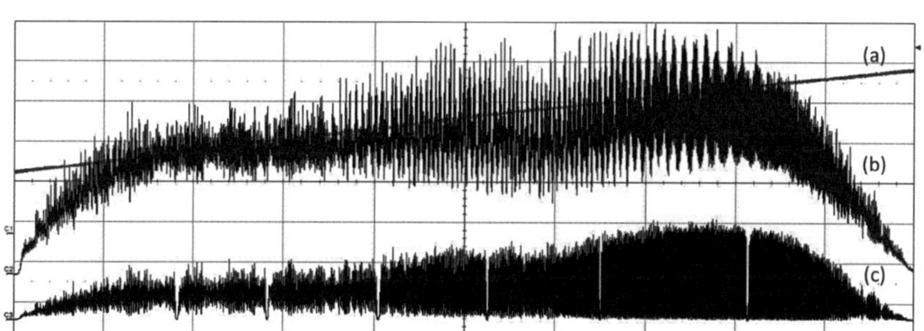

Figure 6.
Typical records obtained with an oscilloscope when measuring the thickness of a sample.

The Fourier transform of this signal can be expressed as indicated in Eq. (14), where it can be seen that the spectral profile of the light source is convolved with the deltas coming from the DC term and the interfering term of Eq. (13):

$$I(x) = S(x) \otimes [\delta(x) + \delta(x \pm (n\, d))] \tag{14}$$

Then, the signal was passed through a high-pass filter that eliminates the DC component, obtaining a signal as indicated in **Figure 8**, which can be expressed as

$$I'(k) = S(k) 2\sqrt{I_1 I_2} \cos(k\,(n\,d)) \tag{15}$$

It is evident that the amplitude depends on the spectral profile of the light source ($S(k)$) and it is desirable to eliminate this dependence. For this, the Hilbert transform [42, 43] is used, which gives a quadrature signal (Eq. (16)), that is

$$\tilde{I}'(k) = S(k) 2\sqrt{I_1 I_2} \sin(k(nd)) \tag{16}$$

By making the square sum of both, Eq. (17) is obtained:

$$\left(S(k)\, 2\, \sqrt{I_1\, I_2}\, \sin(k(nd))\right)^2 + \left(S(k)\, 2\, \sqrt{I_1\, I_2}\, \cos(k(nd))\right)^2 = \left(S(k)\, 2\, \sqrt{I_1\, I_2}\right)^2 \tag{17}$$

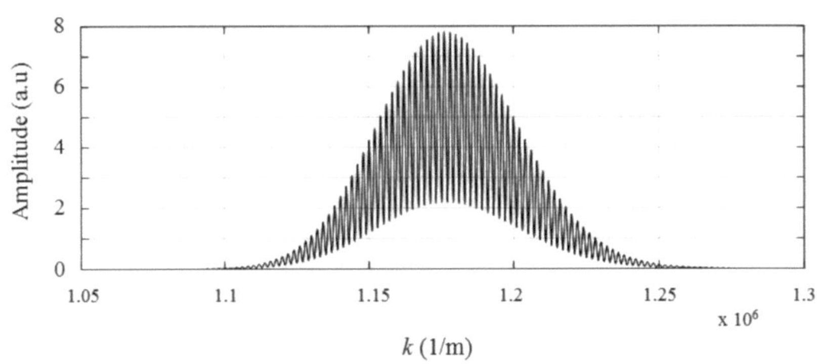

Figure 7.
Interference signal I(k).

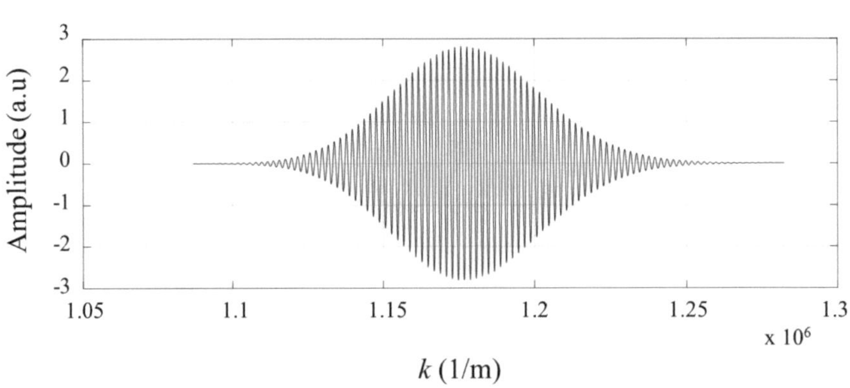

Figure 8.
Interference signal after the high-pass filter.

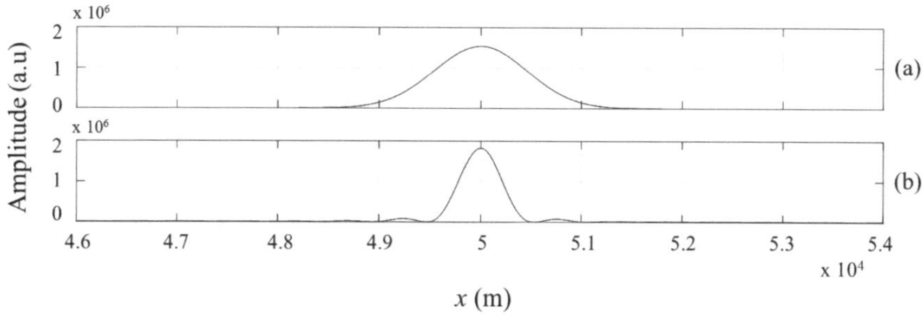

Figure 9.
Fourier transforms of (a) $I'(k)$ and (b) $I''(k)$.

Then, just by calculating the square root of the sum of the squares of the real and imaginary parts, it is possible to eliminate the spectrum of the light source (Eq. (18)):

$$\frac{I'(k)}{S(k)\, 2\, \sqrt{I_1 I_2}} = \cos\left(k(nd)\right) \tag{18}$$

When using this simple method to obtain $S(k)$, there is a problem if it presents zero crossings, which would cause the signal to tend to infinity and produce unwanted results. To avoid this, a constant ξ is usually added to the denominator of the previous expression (Eq. (19)):

$$I''(k) = \frac{I'(k)}{\left(S(k)\, 2\, \sqrt{I_1 I_2} + \xi\right)} = \cos\left(k(nd)\right) \tag{19}$$

Based on this improvement, it is possible to reduce the width at half height by ~40%, improving the uncertainty in detecting the position of the maximum at the peak of the Fourier transform. This improvement is exemplified in **Figure 9**, where graph (a) shows the shape of the peak without performing the Hilbert transform and graph (b) after applying that transform.

Finally, and to improve the S/N ratio, the autocorrelation of $I''(k)$ is performed.

3. Measurements with the OCT system

In this section, different experimental results that demonstrate the potential of the proposed scheme for quality control in industrial applications will be presented [32, 33].

3.1 Long-distance measurements

To apply the proposed system to long-distance measurement, a Michelson-type interferometer was mounted as shown in **Figure 10**. The illumination beam was divided by a 50:50 single-mode fiber coupler, and its outputs were collimated to impinge both on the sample (moving mirror M1) and on the mirror of the reference arm (M2). In order to determine the measuring range of the system, the sample mirror was mounted on a rod with preset positions separated 5 cm, while the reference mirror was placed on a translation stage with a micrometric screw of 5 cm travel. Then, by combining the displacements of both mirrors, it was possible to obtain a continuous variation of the distance to be measured, reaching a possible measuring range of about 20 cm.

Several distances within the measuring range were determined, and it was verified that the system was able to measure up to 17.24 cm. When processing the interference signals using the Hilbert-Fourier transform, it was observed that the peaks lost intensity as the OPD increased. This was due to the divergence of the collimators used in the setup and the coherence length of the light source. The drop in the visibility of the processed signal can be observed in **Figure 11** where, as the

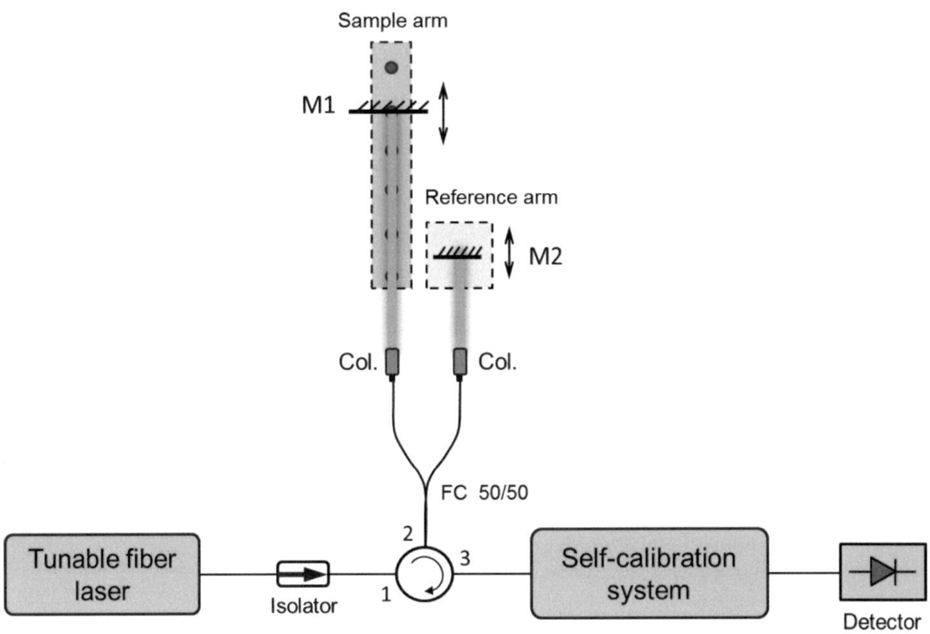

Figure 10.
Experimental setup for long-distance measurement.

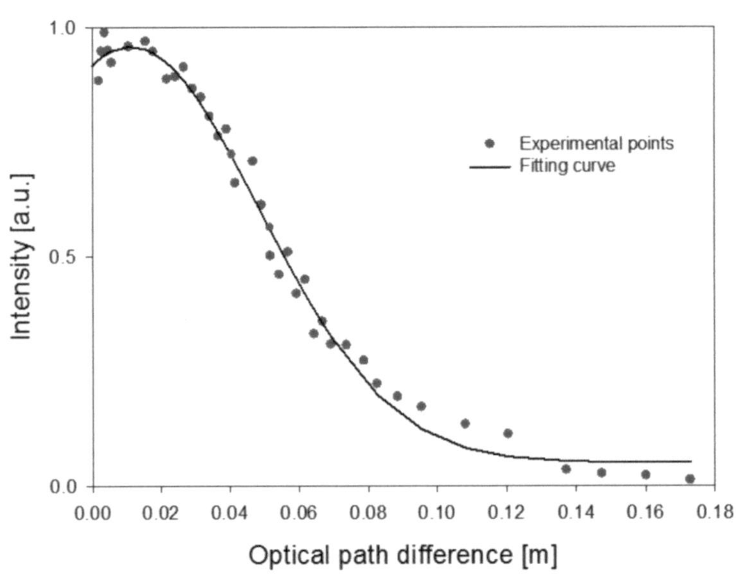

Figure 11.
Normalized amplitude of the Fourier peaks corresponding to different OPDs.

OPD grows, the amplitude of the interference peak decreases, which produces the degradation of the signal-to-noise ratio and limits the maximum measurable distance to the value previously mentioned.

The measurement depth, that is, the theoretical maximum distance that the system can measure (Δz_{max}), is determined by

$$\Delta z_{max} = \frac{\lambda_0^2}{4\, n_g\, \delta_\lambda} \tag{20}$$

where λ_0 is the central wavelength of the source spectrum, n_g is the group refractive index of the medium in which light is propagated, and δ_λ is the spectral with of the tunable laser when it is sweeping in wavelength (somewhat larger than that corresponding to the emission at any fixed wavelength within the tuning range). For the tunable laser source used in our measuring system, it can be considered δ_λ = 30 pm, so, theoretically, Δz_{max} = 20 cm. However, it was experimentally verified that the aforementioned drop in visibility slightly reduced this value to just over 17 cm.

3.2 Measurement of multilayer transparent samples

The system described in this chapter was used to measure the thickness of the walls and distance between them for different transparent and semitransparent containers, of different sizes, colors, and shapes, as well as to dimensionally characterize multilayer transparent objects in order to expand the possible applications.

The setup used corresponds to a single-arm Michelson interferometer. In this type of configuration, the interference signal is obtained from the superposition of the reflections in each of the interfaces present in the sample, which makes it possible to eliminate errors due to reference vibrations and a better use of the light source. Initially, measurements were made on rectangular glass cuvettes and on round glass containers (e.g., ampoules and jars used in the pharmaceutical industry and glass bottles used in the beverage industry), seeking to determine the thicknesses of the walls, the internal and external dimensions or diameters, and the shape in each one. In order to measure cylindrical containers, the samples were mounted on a rotating platform that allows measurements to be taken in different sectors. This type of measurements can be used to perform quality control during the manufacturing process of such containers. In **Figure 12**, a glass jar or cuvette is shown schematically, as well as the dimensions to be measured and the beams reflected in the different interfaces.

If we consider the beams reflected in each interface of the container, 6 interference signals will be generated that will give rise to the corresponding Fourier transform peaks, from which it will be possible to obtain the information of the dimensional parameters of the mentioned container (**Figure 13**). If we call $P_{m,n}$ the position on the OPD axis of the Fourier transform peak corresponding to the interference signal generated by the reflections I_m and I_n and n_i the refractive index of the layer (i) of thickness T_i, then the position of the peak can be expressed as

$$P_{m,n} = \sum_{k=m}^{n-1} T_k n_k \tag{21}$$

If we consider the glass container shown in **Figure 12**, the refractive indexes will be those of glass and air (n_g and n_a = 1, respectively), then

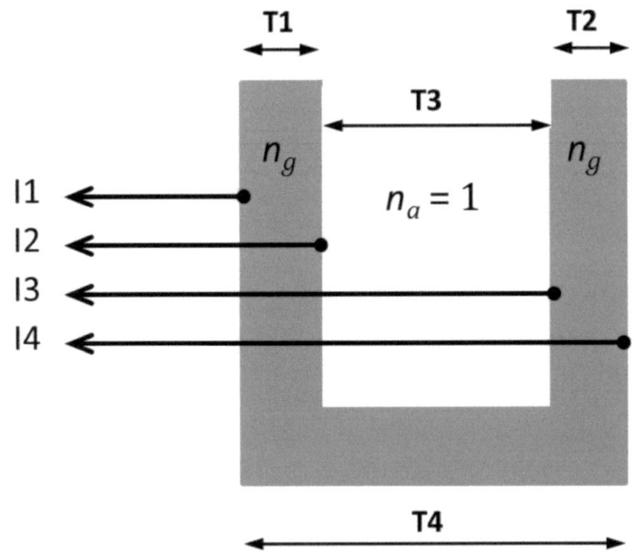

Figure 12.
Generic scheme of a container showing the beams reflected on each interface and the dimensions to be measured.

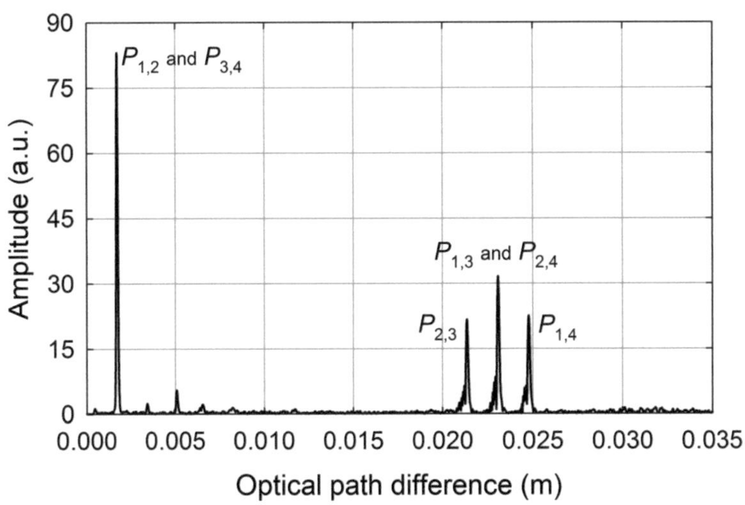

Figure 13.
Fourier transform of the interference signal obtained when measuring a small glass bottle.

$$P_{1,2} = T_1 n_g \tag{22}$$

$$P_{3,4} = T_2 n_g \tag{23}$$

$$P_{2,3} = T_3 \tag{24}$$

$$P_{1,3} = T_1 n_g + T_3 \tag{25}$$

$$P_{2,4} = T_3 + T_2 n_g \tag{26}$$

$$P_{1,4} = T_1 n_g + T_3 + T_2 n_g \tag{27}$$

From these equations it is possible to obtain the dimensions of the container as

$$T_1 = \frac{P_{1,2}}{n_g} \tag{28}$$

$$T_2 = \frac{P_{3,4}}{n_g} \tag{29}$$

$$T_3 = P_{2,3} \tag{30}$$

$$T_4 = P_{2,3} + \frac{(P_{1,2} + P_{3,4})}{n_g} \tag{31}$$

As an example of the different measurements made in this type of samples, **Figure 13** shows results obtained when measuring a small glass bottle.

As can be seen, $P_{1,2}$ was superimposed with $P_{3,4}$, so both thicknesses were equal $T_1 = T_2 = 1.15$ mm; the distance between the walls of the bottle (or internal diameter) arises from $P_{2,3}$ and resulted in $T_3 = 21.3$ mm, while $P_{1,4}$ provides information on the external size (or diameter) of the container, resulting in $T_4 = 23.6$ mm. The peak $P_{1,3}$ coincides with $P_{2,4}$ and comes from the interference between the reflected beam on one of the internal walls and the opposite outer one. **Table 1** summarizes the values obtained after a complete rotation of the same glass bottle.

Other samples like glass ampoules, jars, and beakers like those used in laboratory up to 5 cm external diameter were measured.

Subsequently, in order to expand the possible applications of the developed system, different multilayer samples formed by glass plates separated by layers of air were measured. The objective was to determine the thickness of each of the different layers as well as the total dimensions of the sample. This type of measurement would allow, for example, the determination of the thickness of internal partitions in containers, helping to carry out quality control during the manufacturing processes. Two samples were measured, one consisting of three glass thicknesses and two air thicknesses, and another formed by four glass walls separated by three air spaces. The latter is schematized in **Figure 14**, where the thicknesses of the different layers (G_i and A_i) are indicated, as well as the intensities reflected in the different interfaces (I_i). In this configuration, eight reflections are produced and combined to generate the different peaks in the Fourier transform of the interference signal. **Figure 15** shows the experimental result obtained when measuring this multilayer object. There, only those peaks that provide relevant information to directly determine the searched dimensions were labeled. The other peaks correspond to different combinations between the thicknesses analyzed. Besides, higher amplitude peaks indicate that for this sample there is more than one interface with the same thickness. Simulations performed with the values of the

Angular position	$T_1 = T_2$ (m)	T_3 (m)	T_4 (m)
0°	0.001178	0.021273	0.023598
45°	0.001146	0.021313	0.023601
90°	0.001178	0.021301	0.023644
135°	0.001142	0.021352	0.023639
180°	0.001179	0.021316	0.023642
225°	0.001138	0.021358	0.023656
270°	0.001168	0.021334	0.023675
315°	0.001143	0.021271	0.023554
360°	0.001143	0.021271	0.023554

Table 1.
Measurement of thicknesses (T1 and T2) and internal (T3) and external (T4) diameters of a small glass bottle when rotating on its axis of symmetry.

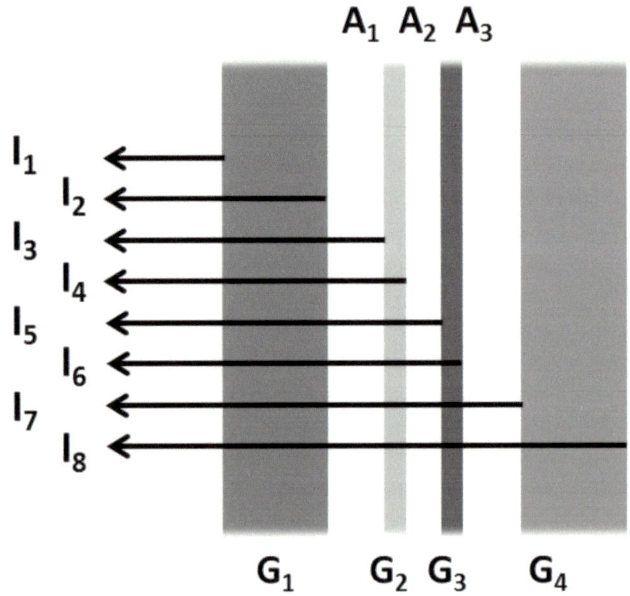

Figure 14.
Multilayer sample consisting of four glass walls separated by layers of air.

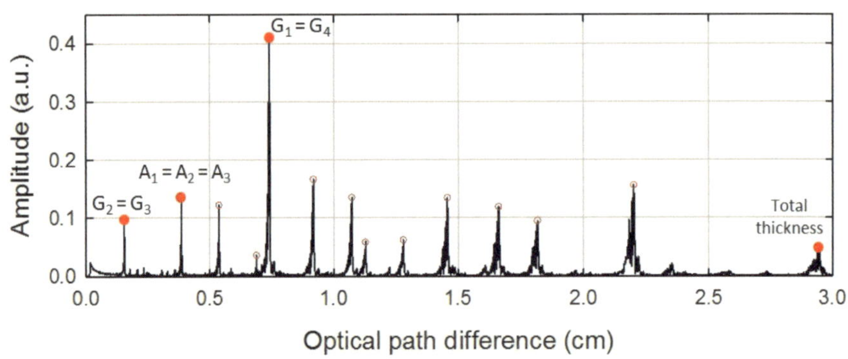

Figure 15.
FFT of the interference signal obtained when measuring a sample of seven layers.

thicknesses mechanically measured showed the same number of interference peaks located in the same spatial positions, so it is evident that this system is capable of characterizing large multilayer samples.

3.3 Measurement of opaque objects

As discussed in Section 2.3.2, the interferometric system was slightly modified to measure opaque samples. The configuration used is shown schematically in **Figure 16**.

Eight opaque samples (aluminum cylinders) of different thicknesses between 5 and 50 mm were measured. For each of the samples, the following were recorded: (1) the interferometric signal obtained with the reference and the Sagnac ring without the object and (2) the interferometric signal with the object placed inside the Sagnac ring and the reference. This allows to obtain the OPDs corresponding to the peaks P_r, P_1, and P_2 of Eq. (12) from which it is possible to determine the

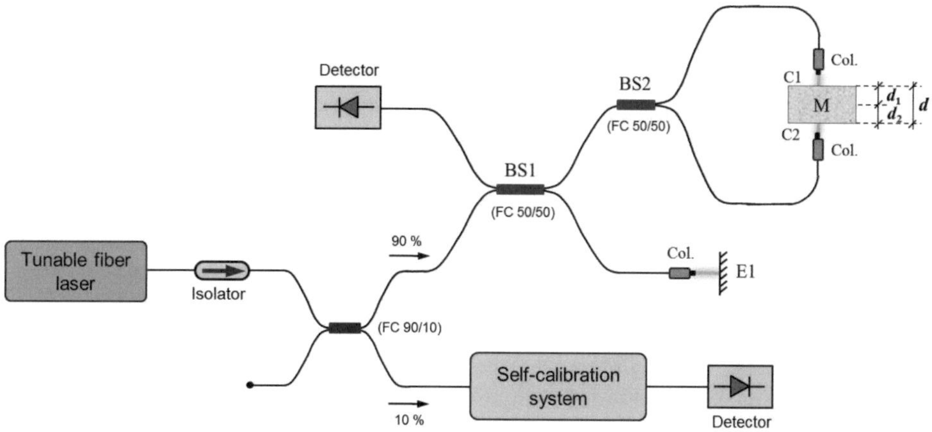

Figure 16.
Experimental setup used to measure opaque samples.

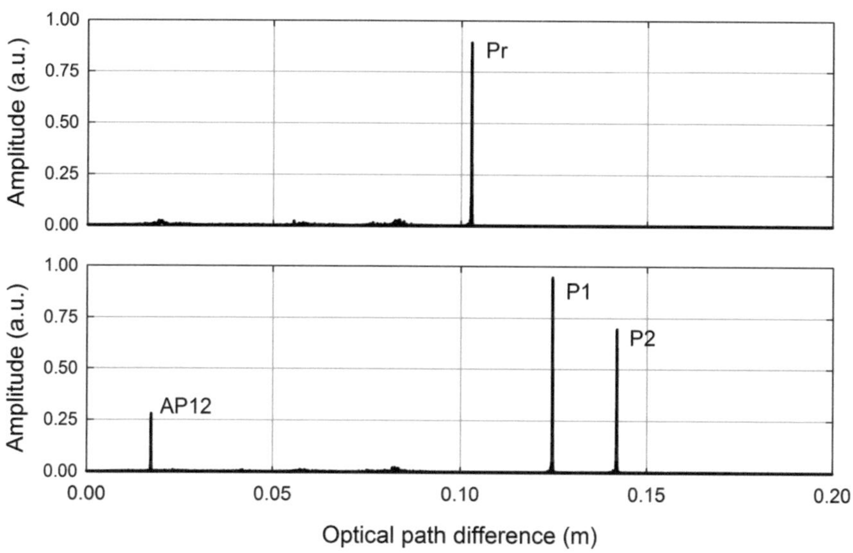

Figure 17.
Fourier transform of the interference signal obtained when measuring an opaque sample of 30.44 mm thickness.

thickness d of the object as explained in Section 2.3.2. In **Figure 17**, the Fourier transforms corresponding to both situations for one of the samples are observed, resulting in a thickness d = 30.44 mm. The AP12 peak always appears in this type of measurements and is the OPD between each of the faces of the sample, that is, it corresponds to the interference signal between the reflections on both sides of the object, although it is not taken into account to determine its thickness.

4. Conclusions

This chapter discusses the development of a simple, self-calibrated system, based on the SS-OCT technique, and its metrological applications in nonmedical areas where the determination of large dimensions (several centimeters) is required. This type of measurement is of great interest, for example, to carry out quality control in manufacturing processes. The developed system allows not only tomography of transparent or semitransparent multilayer samples but also the measurement of thicknesses of opaque mechanical parts. The implemented scheme allowed to measure distances somewhat greater than 17 cm with an axial resolution of 21 μm, which extends the measurement range obtainable with the usual OCT schemes while maintaining a good axial resolution. This is important for applications in areas of metrology and nondestructive testing in the industrial field. The system uses a laser source of ~20 pm bandwidth tunable in the spectral region of 1550 nm, which has sufficiently long coherence length to enable long depth range imaging. The design includes an online, passive self-calibration stage implemented with a set of fiber Bragg gratings. The calibration system provides the correspondence between the wavelength and the temporary location of the laser emission within each sweep, which allows to correct, in a simple way and without additional electronic devices, possible indeterminations in the relation between both parameters. It was verified that the possible temperature changes do not affect the calibration of the system as long as all the fiber gratings experience the same thermal variation. Using schemes as the proposed one, progress can be made in little explored applications of reflectometry, such as in the determination of dimensional parameters of mechanical parts (including profilometry of surfaces or parts) and tomography for the study of transparent and semitransparent materials (glass, plastics, polymers, etc.) of large dimensions.

Acknowledgements

This work has been funded by the Consejo Nacional de Investigaciones Científicas y Técnicas (CONICET) (PIP 112-201101-00397), Facultad de Ingeniería de la Universidad Nacional de la Plata (Proyecto I169), Comisión de Investigaciones Científicas de la Provincia de Buenos Aires CIC (Resoluciones N° 1266/14, 602/16, 195/17), and Facultad Regional Delta, Universidad Tecnológica Nacional (PID 2221), Argentina.

Author details

Nélida A. Russo[1*], Eneas N. Morel[2], Jorge R. Torga[2] and Ricardo Duchowicz[1]

1 Centro de Investigaciones Ópticas (CONICET-CIC), La Plata, Buenos Aires, Argentina

2 Laboratorio de Optoelectrónica y Metrología Aplicada, Facultad Regional Delta, UTN, Campana, Buenos Aires, Argentina

*Address all correspondence to: nelidar@ciop.unlp.edu.ar

References

[1] Huang D, Swanson EA, Lin CP, Schuman JS, Stinson WG, Chang W, et al. Optical coherence tomography. Science. 1991;**254**:1178-1181

[2] Tearney GJ, Boppart SA, Bouma BE, Brezinski ME, Weissman NJ, Southern JF, et al. Scanning single-mode fiber optic catheter-endoscope for optical coherence tomography. Optics Letters. 1996;**21**:543-545

[3] Hee MR, Izatt JA, Swanson EA, Huang D, Schuman JS, Lin CP, et al. Archives of Ophthalmology. 1995;**113**: 325

[4] Fercher AF, Hitzenberger CK, Kamp G, El-Zaiat SY. Measurement of intraocular distances by backscattering spectral interferometry. Optics Communication. 1995;**117**:43-48

[5] Lexer F, Hitzenberger CK, Fercher AF, Kulhavy M. Wavelength-tuning interferometry of intraocular distances. Applied Optics. 1997;**36**:6548-6553

[6] Häusler G, Lindner MW. 'Coherence radar' and 'spectral radar'—New tools for dermatological diagnosis. Journal of Biomedical Optics. 1998;**3**:21-31

[7] Boppart SA, Tearney GJ, Bouma BE, Southern JF, Brezinski ME, Fujimoto JG. Noninvasive assessment of the developing *Xenopus* cardiovascular system using optical coherence tomography. Proceedings of the National Academy of Sciences of the United States of America. 1997;**94**: 4256-4261

[8] Rogowska J, Bryant CM, Brezinski ME. Cartilage thickness measurements from optical coherence tomography. Journal of the Optical Society of America. A. 2003;**20**:357-367

[9] Walecki W, Suchkov V, Van P, Lai K, Pravdivtsev A, Mikhaylov G, et al. Non-contact fast wafer metrology for ultra-thin patterned wafers mounted on grinding and dicing tapes. In: IEEE/CPMT/SEMI 29th International Electronics Manufacturing Technology Symposium. 2004. pp. 323-325

[10] Walecki WJ, Lai K, Pravdivtsev A, Souchkov V, Van P, Azfar T, et al. Low-coherence interferometric absolute distance gauge for study of MEMS structures. Proceedings of SPIE. 2005; **5716**:182-188

[11] Guss G, Bass I, Hackel R, Mailhiot C, Demos SG. High-resolution 3-D imaging of surface damage sites in fused silica with optical coherence tomography. Proceedings of the SPIE 6720; 2007. 67201F

[12] Walecki W, Wei F, Van P, Lai K, Lee T, Lau SH, et al. Interferometric metrology for thin and ultra-thin compound semiconductor structures mounted on insulating carriers. In: CS Mantech Conf. 2004

[13] Markl D, Hannesschlaeger G, Leitner M, Sacher S, Koller D, Khinast J. A device and a method for monitoring a property of a coating of a solid dosage form during a coating process forming the coating of the solid dosage form. GB application 2513581 (A), EP application 2799842 (A1), US application 020140322429 (A1); 2014

[14] Walecki WJ, Szondy F. Fiber optics low-coherence IR interferometry for defense sensors manufacturing. In: Proceeding of SPIE; 7322; 2009. p. 73220K

[15] Dufour ML, Lamouche G, Detalle V, Gauthier B, Sammut P. Low-coherence interferometry, an advanced technique for optical metrology in industry. Insight-Non-Destructive Testing and Condition Monitoring. 2005;**47**:216-219

[16] Dufour ML, Lamouche G, Vergnole S, Gauthier B, Padioleau C, Hewko M, et al. Surface inspection of hard to reach industrial parts using low-coherence interferometry. Proceedings SPIE. 2006; **6343**:63431Z

[17] Song S, Xu J, Wang RK. Long-range and wide field of view optical coherence tomography for in vivo 3D imaging of large volume object based on a kinetic programmable swept source. Biomedical Optics Express. 2016;**7**: 4734-4748

[18] Leitgeb R, Hitzenberger CK, Fercher AF. Performance of Fourier domain vs. time domain optical coherence tomography. Optics Express. 2003;**11**:889-894

[19] Chinn R, Swanson EA, Fujimoto JG. Optical coherence tomography using a frequency-tunable optical source. Optics Letters. 1997;**22**:340-342

[20] Bonesi M, Minneman MP, Ensher J, Zabihian B, Sattmann H, Boschert P, et al. A kinetic all-semiconductor programmable swept source at 1550 nm and 1310 nm with centimeters coherence length. Optics Express. 2014; **22**:2632-2655

[21] Huber R, Wojtkowski M, Fujimoto JG. Fourier domain mode locking (FDML): A new laser operating regime and applications for optical coherence tomography. Optics Express. 2006;**14**: 3225-3237

[22] Insight, Akinetic All-Semiconductor Technology. 2016. Available from: h ttp://www.sweptlaser.com/high-speed-sweeping

[23] Huber R, Wojtkowski M, Taira K, Fujimoto JG, Hsu K. Amplified, frequency swept lasers for frequency domain reflectometry and OCT imaging: Design and scaling principles. Optics Express. 2005;**13**: 3513-3528

[24] Huber R, Adler DC, Fujimoto JG. Buffered Fourier domain mode locking: Unidirectional swept laser sources for optical coherence tomography imaging at 370,000 lines/s. Optics Letters. 2006; **31**:2975-2977

[25] Adler DC, Huber R, Fujimoto JG. Phase-sensitive optical coherence tomography at up to 370,000 lines per second using buffered Fourier domain mode-locked lasers. Optics Letters. 2007;**32**:626-628

[26] Huang SW, Aguirre AD, Huber RA, Adler DC, Fujimoto JG. Swept source optical coherence microscopy using a Fourier domain mode-locked laser. Optics Express. 2007;**15**:6210-6217

[27] Huber R, Adler DC, Srinivasan VJ, Fujimoto JG. Fourier domain mode locking at 1050 nm for ultrahigh-speed optical coherence tomography of the human retina at 236,000 axial scans per second. Optics Letters. 2007;**32**: 2049-2051

[28] Iiyama K, Wang LT, Hayashi KI. Linearizing optical frequency sweep of a laser diode for FMCW reflectometry. Journal of Lightwave Technology. 1996; **14**:173-178

[29] Grulkowski I, Liu JJ, Potsaid B, Jayaraman V, Jiang J, Fujimoto JG, et al. High-precision, high-accuracy ultralong range swept-source optical coherence tomography using vertical cavity surface emitting laser light source. Optics Letters. 2013;**38**:673-675

[30] Eigenwillig CM, Biedermann BR, Palte G, Huber R. K-space linear Fourier domain mode locked laser and applications for optical coherence tomography. Optics Express. 2008;**16**: 8916-8937

[31] Bradu A, Neagu L, Podoleanu A. Extra long imaging range swept source optical coherence tomography using re-

circulation loops. Optics Express. 2010; **18**:25361-25370

[32] Morel EN, Russo NA, Torga JR, Duchowicz R. Interferometric system based on swept source optical coherence tomography scheme applied to the measurement of distances of industrial interest. Optical Engineering. 2016;**55**: 014105

[33] Morel EN, Russo NA, Torga JR, Duchowicz R. Application of a long-range swept source optical coherence tomography-based scheme for dimensional characterization of multilayer transparent objects. Optical Engineering. 2017;**56**:084102

[34] Cerrotta S, Morel EN, Torga JR, Alvira FC. Feasibility and analysis of a system of two in tandem interferometers for optical coherence tomography. In: XVII Workshop on Information Processing and Control (RPIC). 2017. ISSN: 0018-9219

[35] Antonacci J, Morel EN, Torga JR, Duchowicz R. Spectral sensor resolution measurement improvements by temporal analysis. In: XVII Workshop on Information Processing and Control (RPIC). 2017 ISSN: 0018-9219

[36] Sallese MD, Morel EN, Tabla PM, Torga JR. Reduction of measurement errors in OCT scanning. Proceeding of SPIE. 2018. ISSN: 1605-7422-e-ISSN: 2410-9045

[37] Cerrotta S, Morel EN, Alvira FC, Torga JR. Large axial range frequency-domain optical low coherence interferometry. IEEE Photonics Technology Letters. 2019;**31**:125-128

[38] Giordana A, Duchowicz R. Development and analysis of a simple tunable erbium ring laser. Proceedings of SPIE. 2011;**8011**:80115A

[39] Kashyap R. Fiber Bragg Gratings. 2nd ed. USA: Academic Press—Elsevier; 2009

[40] Othonos A, Kalli K. Fiber Bragg Gratings: Fundamentals and Applications in Telecommunications and Sensing. Artech House: USA; 1999

[41] Yun SH, Richardson DJ, Kim BY. Interrogation of fiber grating sensor arrays with a wavelength-swept fiber laser. Optics Letters. 1998;**23**:843-845

[42] Oppenheim AV, Schafer RW. Discrete-Time Signal Processing. 2nd ed. Upper Saddle River, New Jersey, USA: Prentice-Hall; 1998

[43] Na J, Choi WJ, Choi ES, Ryu SY, Lee BH. Image restoration method based on Hilbert transform for full-field optical coherence tomography. Applied Optics. 2008;**47**:459-466

Chapter 4

Low Cost Open-Source OCT Using Undergraduate Lab Components

Hari Nandakumar and Shailesh Srivastava

Abstract

Instrument cost is one of the factors limiting the adoption of optical coherence tomography (OCT) from a wider range of applications. We present a couple of OCT devices using optical components which are commonly found in undergraduate-level optics laboratories. These low-cost devices have lower signal-to-noise ratios (SNR) than top-of-the-line commercial offerings, yet can serve most of the needs of academic laboratories. A time-domain full-field (TD-FF-) OCT device has been assembled with Arduino control, which yields sub-4-μm axial and lateral resolutions. This device is useful where quick sample acquisition is not critical, but high resolution is paramount, for example with samples from material-science, or ex-vivo stabilized biological samples. Next, we discuss a spectral-domain (SD-) OCT device which delivers real-time video rate B-scans. This device is useful where real-time signal acquisition is desirable, for example with in-vivo biological samples. Cross-platform open-source software control for both these devices is also made available.

Keywords: full-field optical coherence tomography, spectral-domain OCT, low-cost, open-source, CMOS camera

1. Introduction

Non-clinical uses of optical coherence tomography (OCT) has mainly been in industrial scale non-destructive materials testing, and for the study of invaluable historical artifacts [1]. Purchase costs of commonly available commercial OCT devices remained above five figures in US dollars till recently [2]. The lowest priced OCT device available for purchase is, as of 2019, the OQ Labscope from Lumedica, at US$9995 [3, 4]. The cost of the instrument makes it inaccessible to undergraduate students, especially in developing countries. In this chapter, we present a couple of schemes developed in our lab, utilizing commonly available optics lab components to construct capable OCT instruments—in time-domain (TD-OCT) and spectral-domain (SD-OCT) configurations. Such instruments could then be used to study a wide variety of biological and inorganic samples, by choosing an illumination wavelength in which the sample is relatively transparent or translucent.

IntechOpen

2. Full-field TD-OCT

2.1 Motivation

Full-field optical coherence tomography (FF-OCT) or optical coherence microscopy (OCM) is commonly constructed in a Linnik or Michelson interferometer configuration—that is, with or without microscope objectives—as shown in **Figure 1**. FF-OCT employs a 2-D image sensor like a CCD or CMOS camera as the detector, and en-face images of the sample are acquired at different depths by coherence gating. High axial and lateral resolutions are possible with FF-OCT. High-speed cameras also allow FF-OCT to be employed with swept-sources. But FF-OCT can also be implemented [5] using existing lab components with a minimum of parts being purchased. This approach assumes static samples, since the acquisition time for an OCT volume can take a few minutes.

The use of FF-OCT in the time-domain configuration avoids specialized and potentially expensive components such as galvanometer or MEMS micromirror scanners, swept-sources and even superluminescent diode (SLED) or halogen or supercontinuum sources. Simple LEDs provide sufficient illumination for sub-10-μm imaging. Such a system has the potential to create high resolution 3D tomograms at very low cost. The system schematic and an illustrative photograph are shown in **Figure 1**.

2.2 Choosing the right components

2.2.1 Source and detector

The axial resolution of the FF-OCT device directly depends on the choice of light source used, since axial resolution [1] is given by

$$\Delta z = \frac{2\ln 2}{\pi}\frac{\lambda^2}{\Delta\lambda}. \tag{1}$$

where λ is the center wavelength of the source and $\Delta\lambda$ is the full width at half maximum (FWHM) bandwidth of the source spectrum. Additionally, the detector

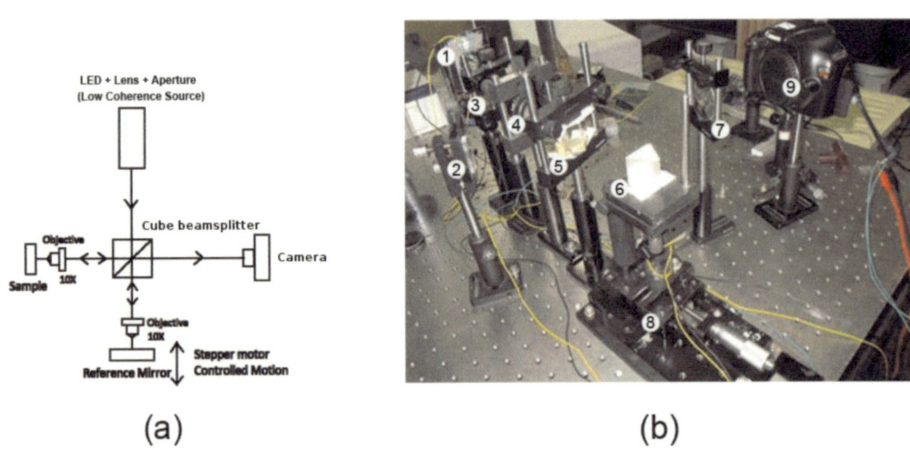

Figure 1.
(a) Schematic of an FF-OCT system with microscope objectives and (b) image of our set-up without microscope objectives. Labeled components are: 1, LED; 2, sample; 3, lens; 4, aperture; 5, cube beam splitter; 6, front surface of a prism as reference mirror; 7, lens; 8, translation stage actuated by stepper motor; 9, camera.

or camera sensor should also have adequate sensitivity over the source spectrum. If the source spectrum is broader than the camera's sensitive region, only the part of the source spectrum convolved with the camera's spectral sensitivity would be the relevant source bandwidth in the Eq. 1 for axial resolution. Thus, if a standard RGB CMOS or CCD camera is used, only the individual red, green or blue sensor bandwidth would be relevant. Hence, it is advisable to use a monochrome camera if submicron axial resolutions are desired. Additionally, consumer-grade cameras almost invariably have IR filters attached to their sensors. These filters will severely limit the use of such cameras with near-IR sources.

Now for some real-world numbers relating to center wavelengths, spectral bandwidths and achievable axial resolutions. The source spectrum of a white LED is shown in **Figure 2**. Though this spectrum is far from the Gaussian spectral shape assumed in Eq. 1, we can assume a spectrum of 200 nm bandwidth centered at 540 nm based on the full-width at half-maximum (FWHM) from the weighted center of the spectrum. With such a source, Eq. 1 would let us have a best-case axial resolution of 0.64 μm if a monochrome camera is used. A digital SLR camera, Canon 1100D, used as the detector for a white LED source resulted [5] in sub-4 μm axial resolutions, due to the convolution of the source spectrum with the RGB sensors'

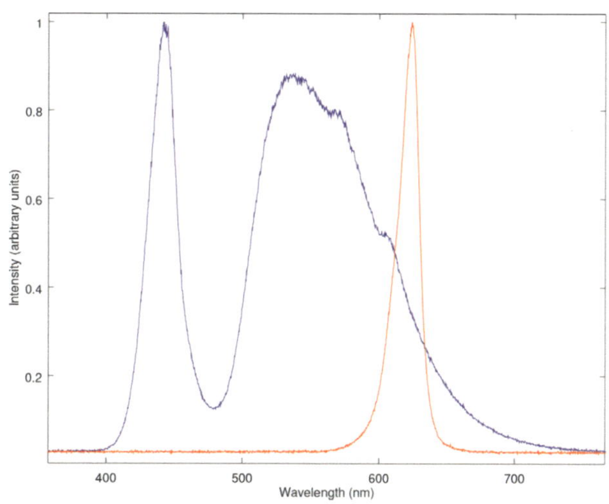

Figure 2.
Source spectra of LEDs obtained using an Ocean Optics USB 4000 spectrometer: white LED in blue, red LED in red.

Figure 3.
An example of a low-cost monochrome CMOS camera: QHY5L-II M.

sensitivity discussed above. In this case, the R data alone was used, thus filtering out the G and B parts of the spectrum and reducing the available source bandwidth. Dispersion compensation is also critically important when broad spectrum sources are used. We discuss dispersion compensation in more detail in Section 2.3. The spectrum of a red LED is shown in **Figure 2**. Using its FWHM of 20 nm and center wavelength of 670 nm, Eq. 1 yields an axial resolution of 9.90 μm. This is quite adequate for many purposes. A camera without an IR filter, like the monochrome CMOS planetary imager (QHY5L-II M) shown in **Figure 3**, enables the use of near-IR sources. Longer wavelengths like near-IR have greater penetration depths in biological samples which exhibit strong Rayleigh scattering. IR LEDs are commonly available with center wavelengths of 850 nm and FWHM of 25 nm, which can give us an axial resolution of 12.8 μm.

2.2.2 Optical components

Uncoated components can be used in the experimental setup for lowering cost. Optical losses due to component reflections would need to be compensated by increasing exposure time at the camera. A cube beam splitter is used in order to avoid dispersion mismatch between the interferometer arms. In case a plate beam splitter is used, dispersion compensation should be done by the insertion of a glass plate of similar thickness in the shorter arm as explained in Section 2.3. The lateral resolution of the system directly depends on the quality of the lenses used. The expression for the lateral resolution of an OCT system [1] is similar to that of a microscope,

$$\Delta r = \frac{4\lambda f}{\pi D} \tag{2}$$

where f and D are the focal length and aperture size of the objective lens. Commonly available microscope objective lenses are listed in terms of magnification available for a standard length microscope, like 10×. A typical 10× objective like the Newport M-10× [6] has a focal length of 16.5 mm and a clear aperture of 7.5 mm. Theoretically, this should yield an axial resolution of 1.56 μm. Experimentally, imaging with uncoated lab lenses uncorrected for aberrations, resolution below 4 μm [5] was obtained.

2.2.3 Automated signal acquisition and processing

Standard single axis translation stages like the TS-35 from Holmarc can be converted to micro-stepping operation using 3D-printed stepper motor coupling shafts which are commonly available online. Driving stepper motors with microstepping drivers like the Bholanath BH MSD 2A enable sub-micron translations. The microstepping driver just needs a TTL signal to initiate the microstep. This can be automated using microcontroller platforms like Arduino. TTL outputs can also be generated by PCs using serial or parallel ports. The sample is attached to the translation stage and moved axially through the microscope objective's focus which should coincide with the zero-path-difference point. Simple automation scripts can then be written to acquire images and translate the sample in the axial direction. We have open-sourced Arduino and OpenCV-based code [7] to acquire and process FF-OCT volumes using the camera SDK from QHYCCD. This code can be easily modified to work with SDKs and cameras from other manufacturers, or even with simple webcams. A screenshot of this software is shown in **Figure 4**.

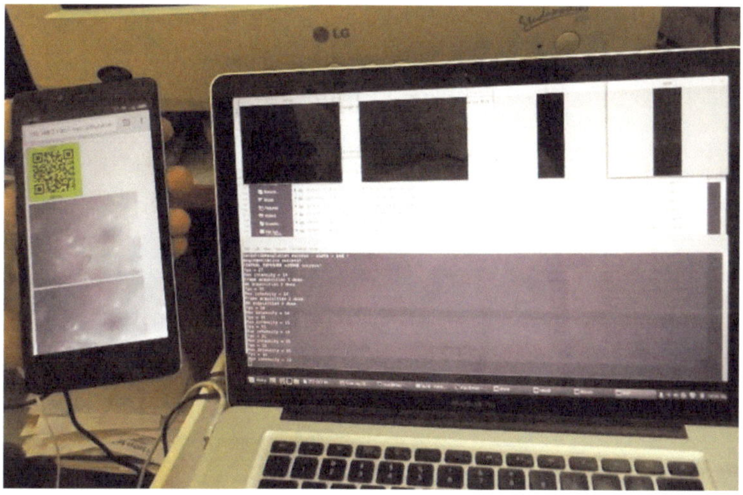

Figure 4.
Screenshot of open-source software for acquiring FF-OCT images.

2.3 Experimental method

The first hurdle in assembling a low-coherence interferometer is accurate estimation of the zero-path-difference point in the two interferometer arms. Our method to do this was as follows. We first verify interferometer alignment by observing coarse interference fringes using visible laser illumination, making sure the fringes remain similar in shape when the moving arm is translated. Next, we translate the moving arm 10 μm at a time under narrow-spectrum LED illumination, monitoring the camera output continuously until interference fringes are visible. Once the zero-path-difference point is identified, the illumination can be changed to wide-spectrum white LEDs or halogen if desired. If the sample is being imaged through a medium, for example if the sample is being used with an oil-immersion objective lens, high axial resolutions with broad-spectrum sources are possible only with dispersion correction. This can be done experimentally by introducing glass plates or cover slips of varying thickness to minimize the axial displacement over which fringes appear for a particular reflective layer.

Many inexpensive camera sensors including mobile phone cameras now allow saving of images in RAW format. Saving in RAW format allows the export of images with a gamma of 1.0, i.e., having a linear detector response to incident optical power. Software like DCRAW can then be used to convert the RAW images to linear TIFF [8, 9] or other image formats. Acquisition of RAW images is not mandatory. If a simple webcam is used as the detector, its output may not be linear with optical power, but interference fringes would still be visible. Such images also could be used to create tomograms, with the rider that the intensities of the reflectors may not be accurately portrayed. Qualitative imaging can still be done, and feature positions would not be affected by the nonlinear camera response.

Post-processing of the images involves subtracting the background DC and taking the Hilbert transform [1] of the images (perpendicular to the fringe lines). The resulting en-face tomographic images can either be volume rendered using free software like Blender [10, 11] or Octave [12], or displayed as single slices in movies and so on. Some examples of tomograms rendered with Blender and Octave are shown in **Figure 5**. Lateral calibration, pixels per micron, can be done by imaging

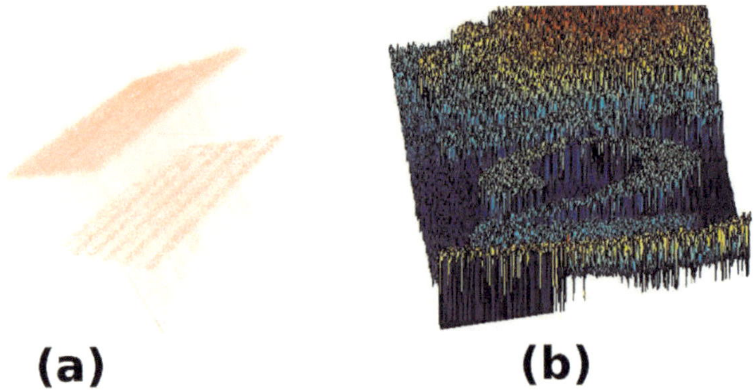

Figure 5.
Tomograms: (a) microchannels in PDMS rendered in blender. (b) Profile of numeral 2 on a coin, plotted in octave.

known samples like USAF targets. Axial calibration directly results from the known motion of the micro-stepping translation stage. The distances we measure with OCT are optical path lengths—actual distances inside samples would be the optical path length divided by the sample refractive index.

3. Parallel SD-OCT

3.1 Motivation

Signal-to-noise ratio (SNR) of Fourier-domain optical coherence tomography (FD-OCT) can be higher by several orders of magnitude compared to time-domain optical coherence tomography (TD-OCT) [13]. This motivates us to devise a way to use the components we have used for FF-OCT in an FD-OCT configuration. FD-OCT can be implemented either with swept-sources (SS-OCT) or using spectrometers as spectral-domain OCT (SD-OCT). Swept-sources are expensive, so we do not attempt to assemble a Swept-Source OCT (SS-OCT) device. Instead, we explore a way to use the two-dimensional camera sensor in a parallel spectrometer design. The schematic of a parallel FD-OCT [14, 15] setup, also known as line-field OCT [16, 17] is shown in **Figure 6**. Such a configuration captures data for a complete B-scan in a single frame captured by the camera. Even a low-cost CMOS camera which can do 200 fps for 320 × 240 frames would then enable an FD-OCT system with a respectable 48,000 A-scans per second.

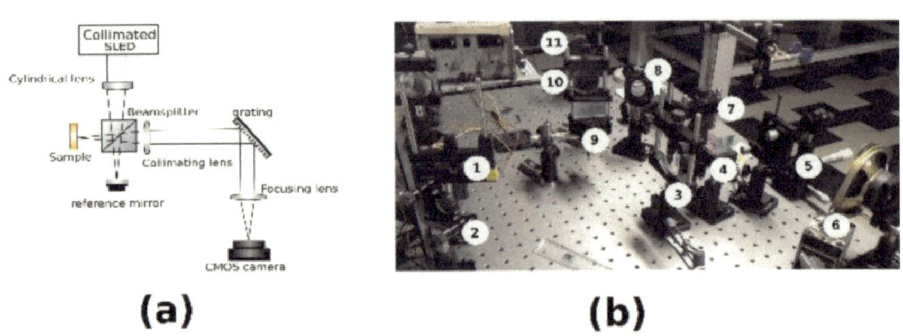

Figure 6.
(a) Schematic of a parallel FD-OCT system (b) our set-up. Labeled components are: 1, collimated SLED source; 2, steering mirror; 3, cylindrical lens; 4, cube beam splitter; 5, plane surface of NDF as sample; 6, reference mirror; 7, collimating lens; 8, steering mirror; 9, grating; 10, imaging lens; 11, camera.

3.2 Specialized components needed

The parallel OCT configuration manages to avoid scanning mirrors, but an SLED source would be needed to focus the line illumination on the sample below a thickness of 100 μm. The extended nature of common LED sources prevent them from being tightly focused in this manner. The $300 SLED (Superlum SLD-340-MP-TO56) and a short focal length aspheric lens for collimation are hence necessary purchases. Care needs to be taken to avoid back-reflections, which can cause damage to SLEDs. The collimating lens should therefore have an anti-reflection coating at the SLED's wavelength range. A diffraction grating or other dispersive element would also be needed.

Here, we would like to point out that SLEDs need greater care in handling than LEDs. Not only are they static-sensitive, they also degrade if overheated or overdriven. Many varieties of free-space SLEDs are also highly sensitive to back-reflections, which can even cause irreversible damage when operated at high powers. Common laboratory power supplies tend to produce large spikes when they are powered on or off, and if an SLED is being driven by such a power supply, it will undergo irreversible damage, reducing its power output. Dedicated driver modules, or laser diode drivers, are the preferred sources to drive SLEDs. The minimum precautions needed would be to employ good heat-sinking, drive at only 50% of rated power, and isolate the SLEDs from power supply spikes using extra switches.

Unlike the TD-OCT case, linear detector response is important for FD-OCT. The noise-floor of the final B-scan also depends on the SNR of the camera. Hence, the camera needs to be chosen with some care, and needs to have gamma, gain and exposure-time controls at the very minimum. We have shown [18] that with simple observations, we can estimate the noise level CMOS cameras. Sensors exhibiting low-noise characteristics along with a high dynamic range are preferable, but useful work can still be done if the camera's SNR is only 40 dB, as with 8-bit cameras which have a dynamic range of only 48 dB. Averaging helps to reduce noise, with the statistical noise reduction of $\sqrt{N}$ for N averages.

3.3 Experimental method

3.3.1 General setup

The cylindrical lens used in the parallel FD-OCT system illuminates a line on the sample (and the reference arm). The spectrum of this line is dispersed by the grating onto the 2D camera sensor. In our implementation, the line is along the vertical axis of the sensor, and the spectrum is thus obtained along each horizontal line of pixels of the camera. Though the post-processing of the captured data is more involved, the practical setting up the FD-OCT assembly is much easier than in the TD-OCT case, since finding the zero-path-difference point is not critically important. Focused width of the line illumination can be measured by imaging the line on the camera. The grating equation points to the angle at which the imaging lens and camera assembly needs to be set up. Once the spectrum is obtained on the camera, the standard FD-OCT post-processing yields a B-scan from each frame.

3.3.2 Spectrometer calibration

In Section 3.4, we discuss the details of the data processing needed to create B-scans from the interograms obtained with the FD-OCT assembly. The required input for the data processing is the calibrated spectrogram, i.e., intensity as a function of wavelength or wavenumber. For this, we need to calibrate the spectrometer consisting of the dispersive element (grating), focusing optics and camera. Since our grating produces a spectrum which is linear in wavelength, we just need to find the bandwidth of the spectrometer for its calibration. This is because we align the camera to have the spectrum centered around the known center wavelength of the source. We used a simple iterative calibration procedure which was easy to use because our software permitted configuration via a text file of parameters. Our calibration procedure was:

1. Enter the source's FWHM bandwidth as an initial estimate of our spectrometer measurement bandwidth, as the min/max wavelengths in our software's ini file.

2. With these settings, use a single surface reflector as the sample in the FD-OCT device. By moving the reflector using a calibrated micrometer translation stage, note the number of B-scan pixels corresponding to a known distance, say 1 mm, of translation. From the known total number of pixels in the B-scan, compute the measurement depth of our OCT device.

3. From FD-OCT theory, the measurement depth of an FD-OCT device, if not corrected for the complex conjugate, [1] is

$$z_{max} = \frac{\lambda_0^2}{4 \cdot \Delta\lambda}. \tag{3}$$

where λ_0 is the source center wavelength and $\Delta\lambda$ is the full width at half maximum (FWHM) bandwidth. If the spectrometer measures a smaller bandwidth than the source bandwidth, the $\Delta\lambda$ would become the spectrometer bandwidth. We use the known value of λ from the source datasheet. Our measured z_{max}, then gives us the required $\Delta\lambda$. We thus calibrate our spectrometer as having a minimum wavelength $\lambda_0 - \Delta\lambda/2$ and a maximum wavelength $\lambda_0 + \Delta\lambda/2$ and enter these parameters in our software's parameter ini file.

3.3.3 Measurable depth and axial resolution

We note that there could be a trade-off between measurable depth and axial resolution, when the number of spectrometer pixels is limited. For example, if a sensor with resolution 1280×960 is used with 2×2 binning, the number of usable spectral points, without interpolation, is just 640. Eliminating the complex conjugate half of the Fourier transform, the resulting B-scan would have only 320 points. If a measurement depth of 6.4 mm is obtained with such a sensor configuration, the number of pixels would limit the axial resolution to 20 μm even if the source characteristics can provide a higher resolution using Eq. 1. We note that the spectrometer resolution can easily be varied by changing the relative powers of the collimating lens and imaging lens in the spectrometer section of the FD-OCT setup. In our setup as seen in **Figure 6**, we have used,

a. cylindrical lens with $f = 15\ cm$

b. cube beamsplitter with NR coating 700-1100 *nm*

c. reflection grating with 1200 *lines/mm* (Richardson gratings 360R25x50x6 from Newport)

d. output collimating lens with $f = 24\ cm$

e. imaging lens with $f = 10\ cm$

and obtained a measurement depth of 6.6 mm with an axial resolution of 20.7 μm, with 2 × 2 binning on 1280 × 960 frames.

3.4 Data processing

In our parallel FD-OCT configuration, each horizontal line on the acquired image corresponds to the spectrum obtained from a single point on the sample. According to the theoretical background [19] of FD-OCT, this spectrum corresponds to an A-scan at that sample point as defined by the following relations. Representing the spectrum as a function of wavenumber k,

$$\begin{aligned} I(k) = {} & S(k) \cdot r_R^2 \\ & +2S(k) r_R \int_{-\infty}^{\infty} r_{s'}(l_s) \cos\left(2k(n_s l_s - l_R)\right) dl_s \\ & +S(k) \left| \int_{-\infty}^{\infty} r_{s'}(l_s) \exp\left[i2k(n_s l_s)\right] dl_s \right|^2 \end{aligned} \tag{4}$$

where $S(k)$ is the source's power spectral density, r_R is the reference arm's amplitude reflectivity, $r_{s'}(l_s)$ is the sample arm point's amplitude reflectivity density of a discrete reflector located at a path length l_s inside the sample and n_s is the sample's refractive index. The first term in the right-hand side of Eq. 4 is the reference intensity term. Subtracting DC from the $I(k)$ gets rid of this term. The second term in Eq. 4 is the desirable one in FD-OCT, and is used to extract $r_{s'}(l_s)$ using the inverse Fourier transform. The third term is the self-interference or autocorrelation term. Autocorrelation causes artifacts in FD-OCT. If sample reflectivity is significantly weaker than the reference arm reflectivity, the autocorrelation term can be safely ignored. Thus, data processing for FD-OCT involves DC subtraction and an inverse Fourier transform on the spectral interferogram represented as a function of wavenumber k.

The captured spectrum obtained by dispersion from a grating is equispaced in wavelength, that is, the raw data we obtain is $I(\lambda)$. This needs to be interpolated to a function $I(k)$ equispaced in k. Basic image processing steps like binning and moving average improve the quality of the image. Further, addition of a larger number of FFT points by interpolation improves final B-scan SNR. Thus, a typical pipeline for data processing FD-OCT data would be:

1. Acquire frames—either of source spectrum or of sample interferogram.

2. Process the frames, optionally, with image processing steps like

    a. binning

    b. averaging

    c. moving average filter

    d. median filter

    e. increasing number of points by zero padding in Fourier domain.

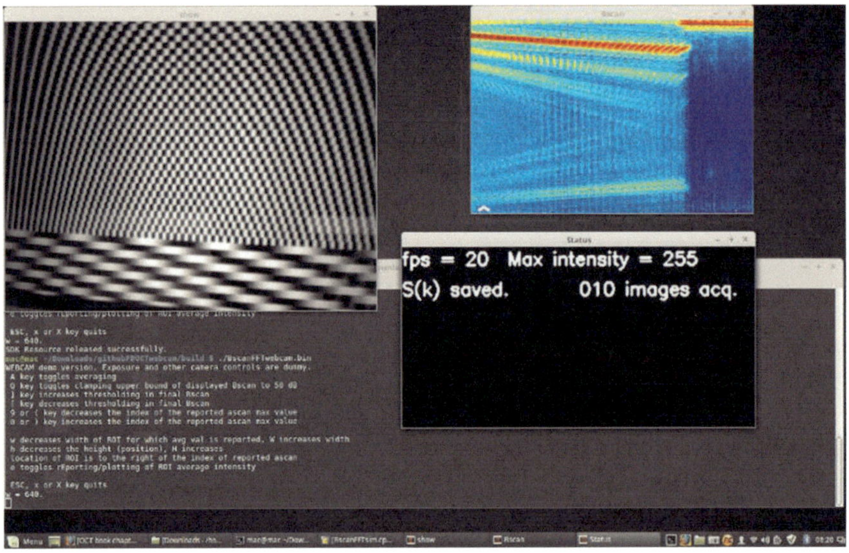

Figure 7.
Screenshot of our open-source software which displays B-scans in realtime.

3. Save a frame with the source spectrum alone.

4. Acquire sample interferogram.

5. Divide the sample interferogram by the source spectrum to obtain the apodized interferogram.

6. Optionally increase the number of points in the interferogram by performing a Fourier transform, zero-padding in the Fourier domain, and then performing an inverse Fourier transform.

7. Resample the interferogram using interpolation to make it equispaced in k rather than equispaced in λ.

8. Perform inverse Fourier transform on the apodized interferogram to obtain the B-scan.

9. Optionally average the B-scan, mask out the DC component, display on screen, save data to disk storage.

We have written real-time code to capture and display B-scans using such a parallel FD-OCT setup. This cross-platform open-source software [20] has been implemented in C++ using the OpenCV [21] framework. Even without using GPU processing, this software can process and display B-scans at 148 frames per second (fps) for low resolution 320 point FFTs, and at 48 fps for 1280 point FFTs, on an i7 MacBook Pro. A screenshot of this software in action is presented in **Figure** 7.

4. Conclusions

In this chapter, we discussed the possibility of assembling capable OCT devices using common laboratory components. Such devices can be put together at low cost, or at no cost if the items are already available, in just a few hours. Such devices would encourage the use of OCT in University optics laboratories, where they could be utilized in inter-disciplinary studies along with Material Science, Chemistry or Biology departments. The microscopic tomograms produced with OCT avoid the expense and preparation required for electron microscope imaging, and also have the advantage of being non-destructive. Tomographic reconstructions of crystals which can show up micron-level imperfections, measurements of biofilm thickness, sample optical dispersion measurements, volume renderings of micro-machined or laser etched parts, or measurements of thick-film deposition are just a few of the myriad applications which can result from the ready availability of OCT in an academic setting. Lowering the cost barrier to entry for OCT imaging is likely to stimulate innovative investigations.

Acknowledgements

This work was not funded by any grant from any funding agency.

Conflict of interest

The authors declare that they have no conflict of interest to report.

Notes/thanks/other declarations

We thank Bhagawan Sri Sathya Sai Baba, the founder Chancellor of our University, for help and guidance during the low-cost OCT development work.

Appendices and Nomenclature

OCT	optical coherence tomography
FF-OCT	full-field optical coherence tomography
FD-OCT	Fourier-domain optical coherence tomography
SD-OCT	spectral-domain optical coherence tomography
TD-OCT	time-domain optical coherence tomography

We encourage all our academic readers to try out constructing OCT devices similar to the ones we have described. You are welcome to contact us in case you run into any practical difficulties.

Author details

Hari Nandakumar* and Shailesh Srivastava
Department of Physics, Sri Sathya Sai Institute of Higher Learning, AP, India

*Address all correspondence to: hari@radiosai.org

References

[1] Drexler W, Fujimoto JG. Optical Coherence Tomography: Technology and Applications. 2nd ed. Switzerland: Springer International Publishing; 2015

[2] Fujimoto J, Swanson E. The development, commercialization, and impact of optical coherence tomography. Investigative Ophthalmology & Visual Science. 2016; **57**(9):OCT1-OCT13. Available from: https://iovs.arvojournals.org/article.aspx?articleid=2536019

[3] OQ LABSCOPE: Lumedica. Lumedia Inc; 2018. Available from: http://www.lumedicasystems.com/oq-labscope

[4] Kim S, Crose M, Eldridge WJ, Cox B, Brown WJ, Wax A. Design and implementation of a low-cost, portable OCT system. Biomedical Optics Express. 2018;**9**(3):1232-1243. Available from: https://www.osapublishing.org/boe/abstract.cfm?uri=boe-9-3-1232

[5] Nandakumar H, Subramania AK, Srivastava S. Sub-4-micron full-field optical coherence tomography on a budget. Sādhanā. 2018;**43**(6):97. Available from: https://www.ias.ac.in/article/fulltext/sadh/043/06/0097

[6] Newport: M-10X. Newport Inc; 2019. https://www.newport.com/p/M-10X

[7] Nandakumar H. ASKlive. GitHub; 2018. https://github.com/hn-88/ASKlive

[8] Johnson MK. Capturing linear images. MIT; 2009. https://web.archive.org/web/20100225203316/http://www.mit.edu/~kimo/blog/linear.html

[9] Coffin D. Decoding raw digital photos in Linux. 1997–2018. http://www.cybercom.net/%7edcoffin/dcraw/

[10] Foundation B. Blender v2.79b; 1998–2018. https://www.blender.org/

[11] Kent BR. 3D Scientific Visualization with Blender. USA: Morgan & Claypool Publishers; 2015

[12] John WE, David Bateman SH, Wehbring R. GNU Octave version 4.0.0 manual: A high-level interactive language for numericalcomputations; 2015. Available from: http://www.gnu.org/software/octave/doc/interpreter

[13] De Boer JF, Cense B, Park BH, Pierce MC, Tearney GJ, Bouma BE. Improved signal-to-noise ratio in spectral-domain compared with time-domain optical coherence tomography. Optics Letters. 2003;**28**(21):2067-2069 Available from: https://www.osapublishing.org/ol/abstract.cfm?uri=ol-28-21-2067

[14] Grajciar B, Pircher M, Fercher AF, Leitgeb RA. Parallel Fourier domain optical coherence tomography for in vivo measurement of the human eye. Optics Express. 2005;**13**(4):1131-1137. Available from: https://www.osapublishing.org/oe/abstract.cfm?uri=oe-13-4-1131

[15] Nandakumar H, Parameshwaran S, Gamini R, Srivastava S. Artifact-free robust single-shot background subtraction for optical coherence tomography. OSA Continuum. 2019; **2**(5):1556-1564. Available from: http://www.osapublishing.org/osac/abstract.cfm?URI=osac-2-5-1556

[16] Lawman S, Dong Y, Williams BM, Romano V, Kaye S, Harding SP, et al. High resolution corneal and single pulse imaging with line field spectral domain optical coherence tomography. Optics Express. 2016;**24**(11):12395-12405. Available from: http://www.opticsexpress.org/abstract.cfm?URI=oe-24-11-12395

[17] Kazaili A, Lawman S, Geraghty B, Eliasy A, Zheng Y, Shen Y, et al.

Line-field optical coherence tomography as a tool for in vitro characterization of corneal biomechanics under physiological pressures. Scientific Reports. 2019;**9**(1): 6321. Available from: https://www.nature.com/articles/s41598-019-42789-4

[18] Nandakumar H, Srivastava S. Data and code for analyzing performance of QHY CMOS cameras. 2018. Available from: https://www.preprints.org/manuscript/201810.0179/v1

[19] Wang LV, Hi W. Biomedical Optics: Principles and Imaging. USA: John Wiley & Sons; 2012

[20] Nandakumar H. FDOCT. GitHub; 2019. Available from: https://github.com/hn-88/FDOCT

[21] Bradski G, Kaehler A. OpenCV. Dr. Dobbs Journal of Software Tools. 2000; **25**:120-125

Section 2

Non-Medical Imaging Applications

Chapter 5

Optical Coherence Tomography for Polymer Film Evaluation

Jianing Yao and Jannick P. Rolland

Abstract

Development of functional polymer films and film stacks has been under increasing demand to create new generations of novel, compact, light-weight optics. Optical coherence tomography (OCT) is capable of evaluating all the important optical properties of a film or film stack, including topology of surfaces or layer-to-layer interfaces, the refractive index and thickness, and polarization property. By engineering the scanning architecture of an OCT system, high-precision metrology of films of either flat or spherical geometry is achieved. In this chapter, the system design, metrology methodologies, and examples of OCT for film metrology are discussed to provide both the knowledge foundation and the engineering perspectives. The advanced film metrology capabilities offered by OCT play a key role in the manufacturing process maturity of newly developed films. Rapid advancement in the field of OCT is foreseen to drive the application toward in-line film metrology and facilitate the rapid growth of innovative films in the industry.

Keywords: optical coherence tomography, optical metrology, nondestructive inspection, volumetric imaging, scanning architecture, films, surface topography, thickness metrology, refractive index metrology

1. Introduction

In recent years, the application of optical coherence tomography has expanded toward material characterization. The low-coherence nature of optical coherence tomography (OCT) leads to its niche in evaluating films ranging from several microns to a few millimeters thick. OCT is capable of evaluating all the important optical properties of a film or film stack, including topology of surfaces or layer-to-layer interfaces, the refractive index and thickness, and polarization property. OCT inspection of films bypasses the pitfalls of conventional laser Fizeau interferometry in testing film samples that have similar curvatures on both sides. The standard laser Fizeau interferometry is prone to spurious fringes due to the highly coherent light source; the artifacts are prominent when measuring a film sample where multi-reflections between top and bottom surfaces generate undesirable interferences that wash out the signal from the surface under test.

In this chapter, the OCT instrumentation tailored for film inspection is first introduced in Section 2. System calibration methods are described in Section 3 to ensure flat-field, distortion-free mapping of film samples. In Section 4, the methods for simultaneous metrology of the refractive index and thickness of films are presented to inspire the readers. In Section 5, some examples are given to showcase the film metrology results achieved with OCT inspection.

2. Instrumentation

2.1 System architecture

OCT utilizes the interference of low-coherence broadband light, typically with wavelengths in the NIR region. Light divided by a beam splitter is sent to a sample arm that contains the sample under test and a reference arm with a reference mirror. Subsequently, back-reflected signals from the two arms are recombined to form interference fringes that are acquired by a detector for further analysis of the sample information. In time-domain OCT, interferometric fringes are generated by axially moving the reference mirror to match the optical path length of back-reflected signals at different depths of the sample. Interference is only observed when the optical path difference (OPD) between the reference and the sample is within the coherence length of the source.

In comparison, the advancement of Fourier-domain OCT (FD-OCT) has enabled high-speed and high-sensitivity depth-resolved three-dimensional (3-D) imaging of samples. FD-OCT utilizes the principle of light interference in the frequency domain, where reflected light from different depths inside a sample relative to the position of the reference reflector is encoded as different modulation frequencies on the detected spectrum at the interferometer exit. Therefore, no translation of the reference mirror is necessary to achieve depth-resolved imaging of a sample. With the increasing OPD between the reference reflector and the sample structure, spectral modulation frequency increases as well. The resulted summation of interference signals (an A-line spectrum, also referred to as a depth scan) contains spectral modulation frequencies contributed by sample structure at multiple depths. By performing a Fourier transform of the interference spectrum captured at the output of the interferometer, the sample reflectivity profile as a function of depth can be recovered, revealing the internal structure of the sample.

Synchronously with the acquisition of A-line spectra, lateral scanning can be performed either by scanning the focusing sample beam or laterally translating the sample itself. By combining the "depth scan" (no axial movement) and one-dimensional lateral scan, a cross-sectional image can be constructed. A 3-D data set can then be obtained by scanning in the orthogonal lateral dimension and combining the resulting, multiple cross-sectional images. The superb spectrally achieved depth discrimination capability of FD-OCT dramatically improves the imaging speed of OCT, allowing 3-D imaging of samples in real time.

Schematic layouts of an FD-OCT system are shown in **Figure 1**. The development of FD-OCT systems to date may be classified into two types as described in the figure. The first type is spectral domain OCT (SD-OCT) [1–3], in which the sample is illuminated by broadband light from a supercontinuum or superluminescent diode (SLD) source, and a spectrometer is used to collect the interference spectrum. The second type is the swept-source OCT (SS-OCT) [4–7], where a frequency-swept source is employed, in which instantaneous quasi-monochromatic light sweeps across a broad spectrum in a rapid cycle. Corresponding to the swept source, a single-element detector such as a photodiode is used to record the time-encoded spectral interference signal. Furthermore, the ability of FD-OCT to collect signals from the entire depth range of a sample during the entire acquisition time, sampled by multiple spectral channels, leads to significant signal-to-noise ratio (SNR) improvement of 20–30 dB over time-domain OCT

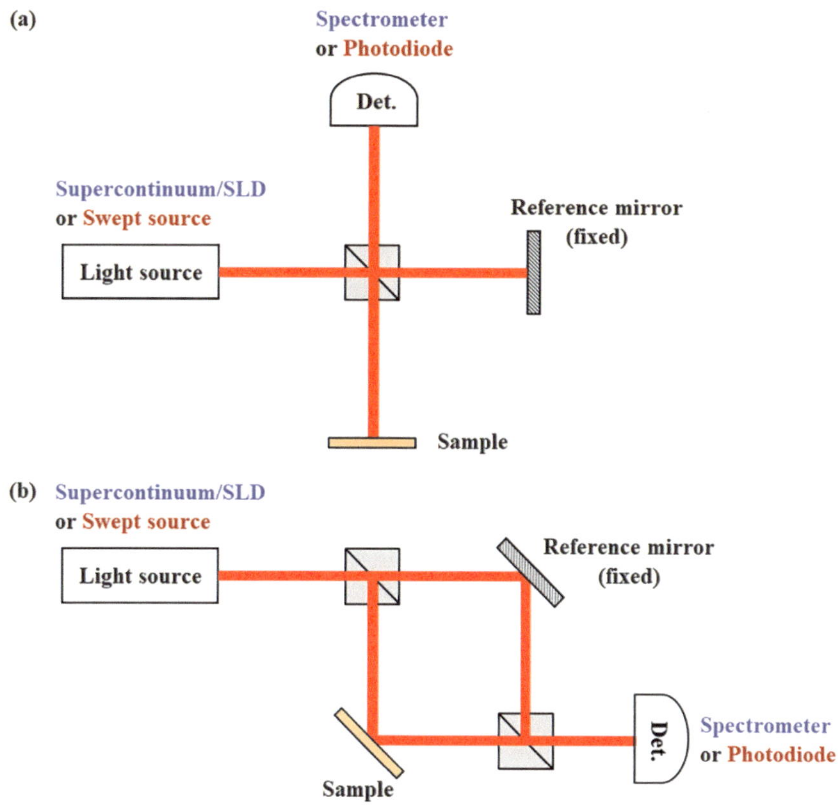

Figure 1.
Schematic layout of typical configurations of an FD-OCT system: (a) Michelson configuration and (b) Mach-Zehnder configuration.

that employs a single-element detector to detect a restricted depth. State-of-the-art OCT systems built nowadays are predominantly FD-OCT.

Figure 1 shows two configurations of an FD-OCT system: (a) Michelson and (b) Mach-Zehnder configuration. It is well known that fringe visibility is maximized when the return power of the sample and reference arms are nearly matched. In the Michelson configuration, the portion of signal attenuation caused by double-pass through the beam splitter is identical for both the sample and reference beams; to generate the maximum amount of interference signal, the beam splitter needs to be at 50/50 split. On the other hand, in the Mach-Zehnder configuration, the power splitting ratio of the beam splitters can be carefully selected based on the optical property of the samples, such that the back-reflected signal of the reference beam balances with that of the sample beam. In this sense, the Mach-Zehnder configuration may allow achieving optimal signal-to-noise ratio at the detector. The ability to boost up sample beam relative to the reference beam is especially important for measuring multi-layer film samples where the back-reflection signals from inter-layer interfaces are weak.

2.2 System design considerations regarding axial point-spread-function and depth range

In OCT systems, the sample is illuminated by broadband light from a supercontinuum or SLD source. Key parameters to consider in designing an OCT system for film inspection are center wavelength, spectral bandwidth, and spectral linewidth, which determine the axial point-spread-function (PSF) and depth range of an OCT system.

Assuming a Gaussian source spectrum, the full-width-at-half-maximum (FWHM) of the axial PSF of an OCT system, denoted as Δz, equals one-half the coherence length of the source as

$$\Delta z = \frac{2\ln 2}{\pi}\frac{{\lambda_0}^2}{\Delta\lambda}. \quad (1)$$

where λ_0 is the center wavelength of the source spectrum and $\Delta\lambda$ is the FWHM of the wavelength spectrum.

It can be seen from Eq. (1) that the FWHM of the axial PSF is proportional to the square of the center wavelength of the light source, and inversely proportional to the bandwidth of the source spectrum. The FWHM of the axial PSF is a common figure of merit used to denote the capability of the system in terms of two-point resolution, which determines the minimal thickness of a film sample that an OCT system can resolve. However, the ability of the system to detect the variation in the axial position of a single reflector well exceeds the FWHM of the axial PSF; it is determined by the sensitivity of the system to sensing even nanometric axial displacements. This axial displacement sensitivity was empirically demonstrated to be often better than one part per thousand of the FWHM of the axial PSF.

It is also worth noting that the axial PSF can be significantly degraded by the dispersion mismatch between the reference and sample arms as a result of the dispersion characteristics of the fiber, sample, and optical components. Therefore, dispersion compensation approaches have been extensively developed by either hardware implementation [8, 9] or numeric correction [10–12].

Moreover, the axial PSF may also be altered by the optical properties of the imaging optics including transmittance, aberrations, polarization, and dispersion. For instance, chromatic aberration of the objective lens would alter the local effective bandwidth and degrade the axial PSF [13]. Therefore, achromatic, all-reflective optical designs may be beneficial for achieving the desired axial PSF of an OCT system.

Additionally, scanning noises, motion artifacts, and detection components may also perturb the axial PSF as well as the sensitivity of the system. All these factors need to be carefully taken into account during the system engineering of OCT instrumentation.

On the other hand, assuming sufficient depth of field provided by the imaging objective lens and an imaging depth range limited by the spectral linewidth that is assumed to be Gaussian in shape, the intrinsic single-sided imaging depth range of an OCT system, denoted as z'_{10dB}, is defined herein as the depth range leading to 10-dB sensitivity roll-off and is computed as

$$z'_{10dB} = \frac{0.895}{\pi}\frac{\lambda_0^2}{\delta_r\lambda}. \quad (2)$$

where $\delta_r\lambda$ is the FWHM of the Gaussian spectral line shape function.

It is worth noting that the sampling interval of the detector needs to be smaller than the instantaneous linewidth of the source in SS-OCT or the spectral resolution of the spectrometer in SD-OCT; otherwise, the usable imaging range will be limited by the sampling.

Finally, when designing an OCT system, a compromise between the axial PSF and imaging depth range usually needs to be considered. As has been shown, both the width of the axial PSF and the imaging depth range are proportional to the square of the center wavelength of the source ${\lambda_0}^2$. Therefore, longer operating wavelengths generally provide larger imaging depth range (although the absorption and scattering coefficients of a sample also need to be considered) at the expense of sacrificing the axial resolution; on the other hand, shifting to lower wavelength regime is expected to bring about better axial discrimination capability at the cost of faster sensitivity decay with depth. Moreover, once the operating wavelength λ_0 is fixed, to push for both a fine axial PSF and a large imaging depth, the challenge arises from the requirement of the system instrumentation to accommodate both a broad spectral bandwidth and a narrow spectral resolution. This would also mean high requirement on the high-speed frequency sweep of the source in SS-OCT, or the large array size of the CCD/CMOS camera in SD-OCT.

2.3 Lateral scanning approaches

FD-OCT enables superb depth sectioning of film samples. To form three-dimensional topography, a lateral scanning mechanism is employed, either by moving the OCT probe or translating the sample in a combination of two orthogonal directions. Different types of lateral scanning architectures are illustrated in **Figure 2**. To measure conventional film samples with flat surfaces, lateral scanning axes are set in Cartesian x and y directions as shown in **Figure 2a–e**; to perform in-situ measurements of curved film layers with spherical surfaces as a result of thermo-molding [14, 15], the lateral scanning axes may be constructed along the azimuthal and polar directions in a spherical coordinate system as shown in **Figure 2f**.

Among all the scanning configurations illustrated in **Figure 2**, **Figure 2e** and **f** move the sample platform by a pair of motorized stages orthogonal in either the Cartesian or spherical coordinates, depending on the curvature of the films under test. 2-D scanning of the sample platform is performed in a straightforward fashion. The straightness, orthogonality, and hysteresis characteristics of the stages are critical to the accuracy of the 3-D data collected. The stages are mechanically tuned to an optimal condition balancing the travel speed, range, and load. Displacement-measuring interferometers are often implemented to provide accurate, real-time feedback of the position of the motorized stages.

Alternative to scanning the sample platform, the beam that is incident on the sample may be sequentially deviated over a sampling grid by changing the angles of the scanning mirrors that are usually located on the path of the beam in the collimated space. Mirror-based scanning of sample beams has gained increasing popularity in OCT systems owing to the high scanning speed it has brought about. Scanning speed up to 500 Hz has been reported due to the low inertia of small mirrors, meaning an entire cross-sectional frame of data may be collected in 2 ms. Popular types of scanning mirrors are galvanometer-based or micro-electro-mechanical (MEMS) type. The common scanning configurations utilizing scanning mirrors in the sample arm of an OCT system are shown in **Figure 2a–c**. As denoted in the diagrams, a collimated beam is reasonably assumed to be the input to the sample arm. In a fiber-based OCT system, the collimated incoming sample beam is commonly formed by a collimating lens placed at one exiting port of a beam-splitting fiber-coupler. In a free-space OCT system, the output from a cube beam splitter dividing the sample and reference beams is collimated. For the simplicity of the illustration, any optics placed before the collimated beam input is omitted in the diagrams. In **Figure 2c**, a single 2-D MEMS scanning mirror may cover scanning field in the entire 2-D plane. In **Figure 2a** and **b**, 2-D scans are achieved by integrating two orthogonal 1-D scans in the x and y directions, respectively.

When scanning the sample beam, preserving telecentric scanning on a sample is crucial to distortion-free mapping of its surface profiles and internal structure, which requires the mirrors scanning in the x and y directions to collocate at the

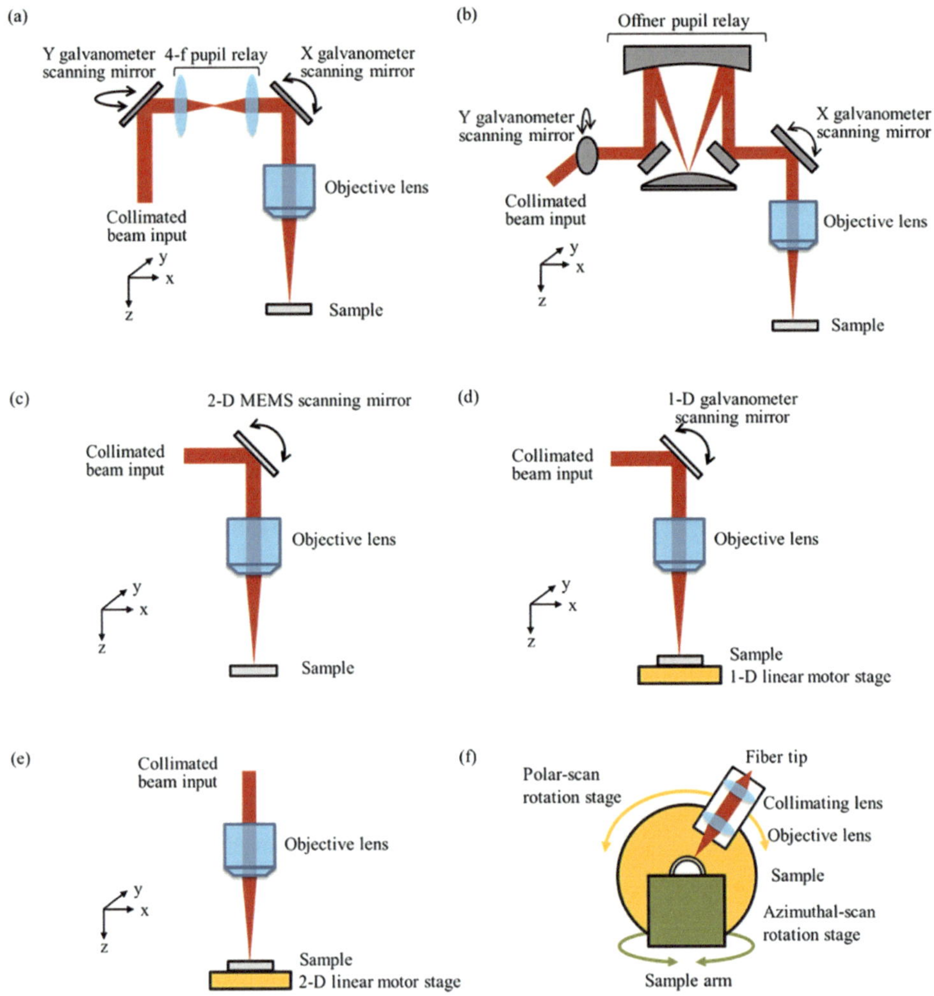

Figure 2.
Illustration of the various lateral scanning configurations in the sample arm of OCT systems.

pupil plane (i.e., focal plane) of the objective lens. Such a scanning geometry is usually not supported by off-the-shelf dual-galvanometer-mirror systems where the pupil plane is simply compromised to be in between the two mirrors that are packaged in a small form factor. While this configuration with the aid of ray-tracing-based software calibration [16, 17] may be suitable for OCT systems toward biomedical imaging applications, it should be avoided in film metrology systems. Any displacement of the pivot of either or both the x-y scanning mirrors from the focal plane of the objective lens would inevitably lead to the deviation of the ray fan from being parallel to the optical axis, and thus introduce optical aberrations including astigmatism, field curvature, and distortion. To ensure the pivot point of both x and y scanning mirrors is at the pupil plane of the objective lens, three viable configurations are shown in **Figure 2a**, **b**, and **d**. **Figure 2d** provides a simple way to bypass the abovementioned pitfall by utilizing a combination of a single axis galvanometer beam steering along the fast-scan axis and a motorized linear stage translating along the orthogonal slow-scan axis. **Figure 2a** and **b** both create another optical conjugate of the pupil plane of the objective lens by implementing a pupil relay optical design. **Figure 2a** employs a refractive 4-f pupil relay design which often consists of a pair of achromatic doublets. **Figure 2b** shows an all-reflective

Offner pupil relay configuration with a concave primary, a convex secondary, and a pair of fold mirrors. The Offner relay carries the benefit of being free from chromatic aberrations across the broad spectral band of an OCT system and provides packaging advantage given its folded geometry.

3. Calibration

To obtain accurate quantitative results from OCT imaging, rigorous calibration of the system needs to be carried out, which generally requires vertical displacement calibration, discussed in Sections 3.1 to 3.3, and lateral scanning calibration, discussed in Section 3.4.

3.1 Spectral nonlinearity calibration

In SD-OCT, raw spectra collected by the CCD or CMOS array of the spectrometer are pseudo linear with the wavelength rather than with the frequency. Similarly, in SS-OCT, the frequency sweep generally is nonlinear with the frequency. In both cases, the detected spectra require calibration to the linear frequency space prior to Fourier transform that converts each spectral signal to a depth profile. The calibration can be achieved either in hardware or software.

In SS-OCT, hardware calibration is performed by using the time-frequency relation measured by an additional side MZI with a fixed optical path difference between its reference and sample arms as denoted by the dashed blue box in **Figure 3**. Simultaneous with the detection of the main interference signal, the calibration signal is detected by a second balanced photo-detector and then digitized by a high-resolution high-speed analog-to-digital converter. The locations of the peaks and valleys of the spectral interference signal from the side MZI that are physically linear in the frequency space provide the resampling scheme needed to recalibrate the spectral interference signal of the main OCT interferometer to be linear in the frequency space. The Fourier transform of a single recalibrated interference spectrum is then performed such that the result shows an entire component reflectivity profile as a function of the linear axial locations along the sample beam path.

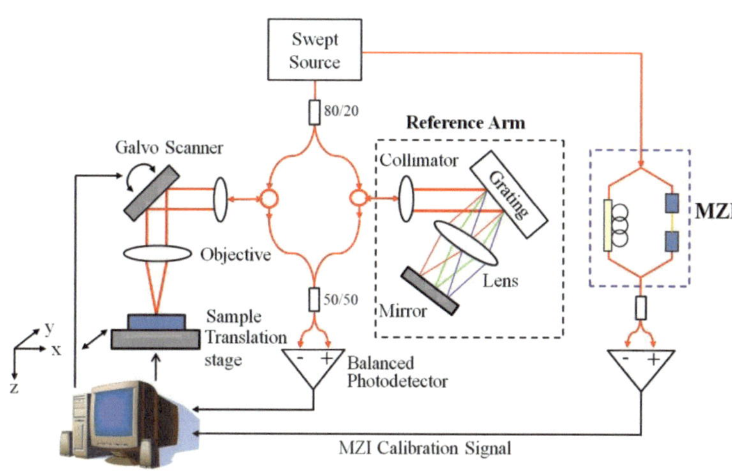

Figure 3.
Layout of an SS-OCT system with a side MZI (denoted in dashed blue box) for spectral nonlinearity calibration.

3.2 Depth scale calibration

After the spectral nonlinearity calibration, rigorous calibration is required to accurately quantify the depth scale on a Fourier-transformed depth profile. Depth scale calibration can be achieved via various methods. A common approach is to image a series of different step height standards that cover the entire z-axis range of interest of the instrument. The details of this calibration approach are described elsewhere [18]. In this section, we describe an alternative method for depth scale calibration.

As show in **Figure 4**, a high surface quality flat mirror is placed in the sample arm where a high-precision motorized linear stage is set up to translate the mirror accurately along the z-axis. A high-resolution displacement measuring interferometer may be implemented to provide accurate feedback of the location of the motorized stage. For the depth scale calibration experiment, the motorized linear stage is controlled to move to a range of z locations covering the entire range of interest along the z-axis. A set of repeated measurements of interference spectra is acquired at each location. Through spectral nonlinearity calibration and subsequent Fourier transforms, depth profiles with a peak showing the mirror signal are restored. The relationship between the physical z positions of the mirror and the average pixel number of the mirror signal detected on the depth profiles are plotted in **Figure 4**. The inverse of the slope of the least squares fit line of the curve represents the depth scale.

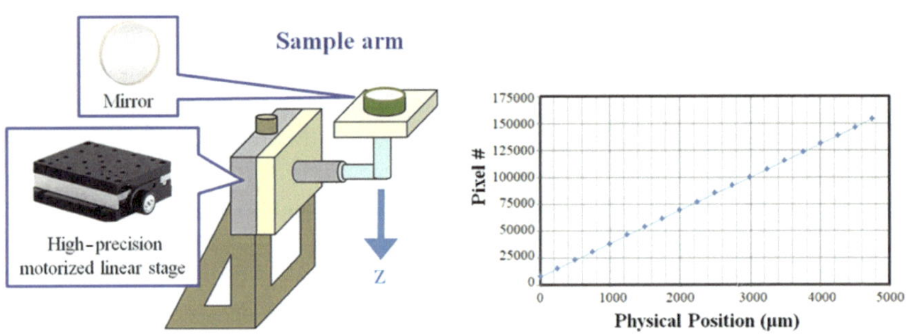

Figure 4.
Illustration of depth scale calibration in OCT systems.

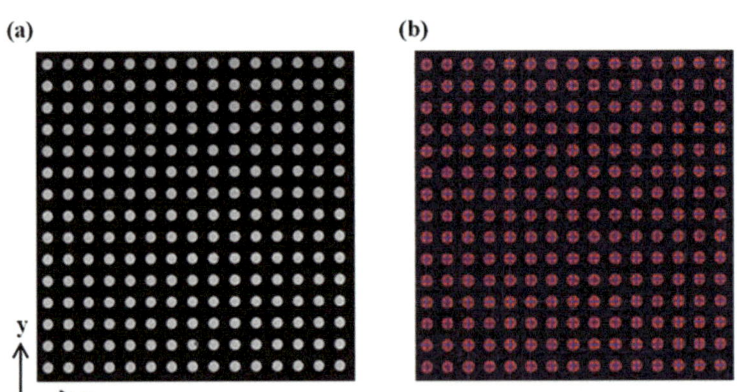

Figure 5.
(a) A raw gray-scale x-y plane image of a dot grid target acquired by an SS-OCT under calibration. (b) a corresponding image after applying a centroiding algorithm to image (a), which shows the detected centroids (blue crosshairs) of the dots overlaying the dot grid (shown as enhanced red dots).

3.3 Field flatness audit

To form accurate 3-D volumetric data, it is important to eliminate any vertical displacement error that may be induced in 2-D scanning of the sample beam. For instance, field curvature may be induced by inaccurate positioning of the x and y scanning mirrors offset from the pupil plane of the objective lens. To inspect the flatness of the x-y scanning field, a high surface quality (e.g., $\lambda/20$) flat mirror may be imaged. Alternatively, a high-grade calibration ball standard may be imaged by angular-scanning OCT system to audit the alignment of the polar scanning axes. These calibration standards are easily measurable on laser Fizeau interferometers to verify the accuracy of the standards and compare against OCT measurement results.

3.4 Lateral scanning field audit and calibration

As described in Section 2.3, various lateral scanning mechanisms are employed in OCT systems to achieve 3-D volumetric imaging. To mitigate scan-induced

Evaluation Criterion	Formula	
Linearity		
Physical dot locations (x, y) vs. pixel coordinates (P_x, P_y)	$\overline{P_x}(x) = \langle f_x(x,y) \rangle_y$,	(3)
	$\overline{P_y}(y) = \langle f_y(x,y) \rangle_x$.	(4)
Orthogonality		
Angle between x and y translation axes with an angle of θ_x and θ_y, respectively, from the horizontal direction of the dot array	$\theta_x = \left\langle \tan^{-1}\left[\frac{\sum_{m=1}^{M}\left(Mf_x(m\cdot dx,y) - \sum_{m=1}^{M} f_x(m\cdot dx,y)\right)\left(Mf_y(m\cdot dx,y) - \sum_{m=1}^{M} f_y(m\cdot dx,y)\right)}{\sum_{m=1}^{M}\left(Mf_x(m\cdot dx,y) - \sum_{m=1}^{M} f_x(m\cdot dx,y)\right)^2}\right]\right\rangle_y$,	(5)
	$\theta_y = \left\langle \tan^{-1}\left[\frac{\sum_{n=1}^{N}\left(Nf_x(x,n\cdot dy) - \sum_{n=1}^{N} f_x(x,n\cdot dy)\right)\left(Nf_y(x,n\cdot dy) - \sum_{n=1}^{N} f_y(x,n\cdot dy)\right)}{\sum_{n=1}^{N}\left(Nf_x(x,n\cdot dy) - \sum_{n=1}^{N} f_x(x,n\cdot dy)\right)^2}\right]\right\rangle_x$.	(6)
Straightness		
Deviation in the perpendicular direction from a true line of travel, residual error of $\varepsilon_x(x)$ for x-travel and $\varepsilon_y(y)$ for y-travel	$\varepsilon_x(x) = < f_y(x,y)\cos\theta_x - x\sin\theta_x - \cos\theta_x \frac{\sum_{m=1}^{M} f_y(m\cdot dx,y)}{M} + \sin\theta_x \frac{\sum_{m=1}^{M} f_x(m\cdot dx,y)}{M} >_y$,	(7)
	$\varepsilon_y(y) = < f_x(x,y)\sin\theta_y - y\cos\theta_y - \sin\theta_y \frac{\sum_{n=1}^{N} f_x(x,n\cdot dy)}{N} + \cos\theta_y \frac{\sum_{n=1}^{N} f_y(x,n\cdot dy)}{N} >_x$.	(8)
Sampling resolution		
Compute x and y pixel size $dP_x(x)$ and $dP_y(y)$ across the travel range	$dP_x(x) = \langle \frac{dx\cos\theta_x}{f_x(x+dx,y) - f_x(x,y)} \rangle_y$,	(9)
	$dP_y(y) = \langle \frac{dy\sin\theta_y}{f_y(x,y+dy) - f_y(x,y)} \rangle_x$.	(10)

Table 1.
List of criteria and formulas used to evaluate the lateral scanning field from imaging a grid target (adapted from [19]).

lateral coordinate errors, the scanning field needs to be calibrated and periodically audited. The typical standards used for this purpose are dot grid targets. The size of the target covers the entire lateral field-of-view (FOV) of the OCT system.

Figure 5a shows an example of a raw gray-scale x-y plane image of a dot grid target acquired by an SS-OCT system with 2-D x-y stages for lateral scanning [19]. From the imaging, a centroiding algorithm is applied to locate the centers of the dots as shown in **Figure 5b**, where the dot grid enhanced in red is overlaid with the detected centroids in blue crosshairs. The two 2-D matrices storing the respective estimated horizontal and vertical pixel coordinates of the detected grid of centroids consist of M (column) $\times$ N (row) elements, where M and N are the numbers of columns and rows of dots imaged, respectively.

Denoting the column and row indices of a dot as (m, n), the nominal physical centroid location of the dot is described as $x = m{\cdot}dx$ and $y = n{\cdot}dy$, assuming dx and dy are constants representing the nominal horizontal and vertical separation between neighboring dots. On the other hand, the measured centroid location from the OCT image can be expressed in pixel coordinates as (P_x, P_y). The mapping from (x, y) to (P_x, P_y) reflects the lateral scanning characteristics of the system and is expressed as $P_x = f_x(x,y)$ and $P_y = f_y(x,y)$. By analyzing the functions f_x and f_y, the attributes of the lateral scanning field consisting of linearity, orthogonality, straightness, and the sampling resolution are estimated as listed in **Table 1**. In the formulas, the notation of angle bracket $< >$ with a subscript x or y denotes averaging over the horizontal or vertical direction of the dot array, respectively, to evaluate the mean (i.e., accuracy) and standard deviation (i.e., repeatability). From the metrics being evaluated from a dot grid target, calibration can be applied to correct the non-orthogonality of the two scanning axes or calibrate the distortion of the field.

4. Methodology for film thickness and refractive index metrology with OCT

To determine the refractive index of a sample, interferometric metrology techniques are generally confronted with the challenge that the measurand obtained reflects the optical thickness of a sample, which is a product of the refractive index and the physical thickness. Using low-coherence interferometry, several techniques have been proposed to decouple the index and physical thickness, which may be categorized into two main methodologies [20]: 1) evaluating the optical distortion of a reference induced by the sample inserted in the path [21–23] and 2) combining low-coherence interferometry with a focus tracking modality such as confocal scanning or multi-photon microscopy [24–27] to obtain another independent relation between index and thickness.

4.1 Simultaneous film thickness and refractive index metrology by evaluating optical distortion induced by a film sample in the sample path

The first method is based on evaluating the optical path length change induced by a film sample inserted in the path. This method is well suited for estimating the refractive index and thickness of a single-layer film using two types of measurements with the same OCT system. It does not require instrumentation modification and is readily adaptable to almost any OCT systems.

Figure 6a and **b** show a schematic cross-sectional view of the experimental setup in the sample arm. In the first step as shown in **Figure 6a**, the film under test is placed above a high-quality reference flat, with a small air gap in between, which is created by a thin ring spacer that supports the film sample in the periphery. Both

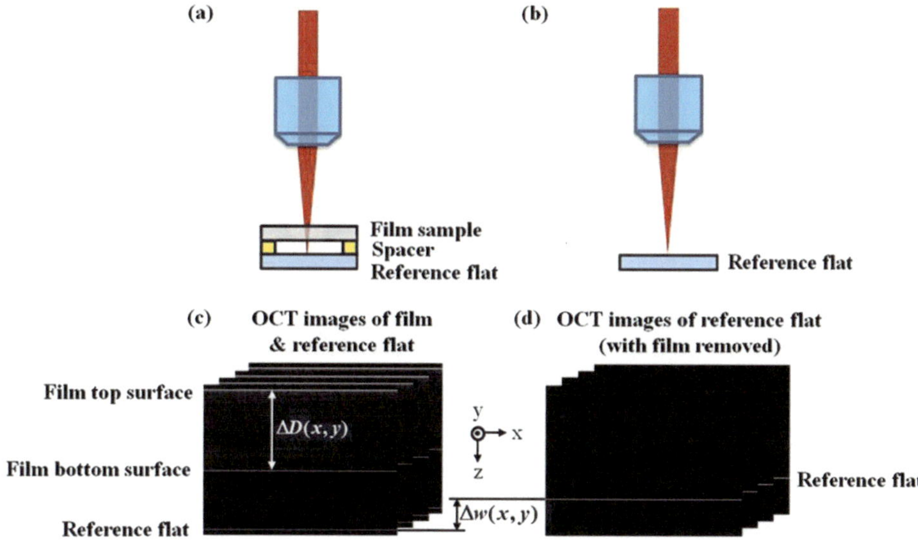

Figure 6.
In the OCT sample arm, (a) a film sample and a reference flat are imaged together, and (b) the reference flat alone is imaged after the film sample has been removed. (c) and (d) are examples of cross-sectional images collected based on setup (a) and (b), respectively.

the film under test and the reference flat need to be properly aligned to ensure normal incidence of the beam on the surfaces being imaged. **Figure 6c** shows an example of raw cross-sectional OCT images collected in this setup. The optical thickness $\Delta D(x,y)$ of the film is obtained by computing the difference between the axial positions of the intensity peaks corresponding to its top and bottom surfaces, which can be expressed as

$$\Delta D(x,y) = \langle n_g(x,y) \rangle_z \cdot t(x,y). \tag{11}$$

where $\langle n_g(x,y) \rangle_z$ denotes the group refractive index averaged over the thickness of the film, and $t(x,y)$ is the physical thickness of the film, both mapped out laterally as a function of the Cartesian (x,y)coordinates. $\langle n_g(x,y) \rangle_z$ may be mathematically expressed as

$$\langle n_g(x,y) \rangle_z = \frac{\int_{z_0}^{z_1} n_g(x,y,z)dz}{t(x,y)}, \tag{12}$$

where $n_g(x,y,z)$ is the local group refractive index within the 3-D volume of a film.

Subsequently, the film sample is removed and only the reference flat was imaged again as shown in **Figure 6b**. An example of raw cross-sectional OCT images is shown in **Figure 6d**. As a result of the previous path of the film sample being replaced by air, the axial position of the reference flat on the OCT image is shifted upward by $\Delta w(x,y)$ due to its reduced optical path difference relative to the reference arm, which can be expressed as

$$\Delta w(x,y) = \left(\langle n_g(x,y) \rangle_z - n_{air} \right) \cdot t(x,y), \tag{13}$$

where n_{air} is the group refractive index of the air at the OCT operating wavelength under laboratory temperature, pressure, and humidity conditions. $\Delta w(x,y)$ reveals the change in the optical path length induced by the film sample. It can be seen from Eqs. (12) and (13) that, by measuring both the $\Delta D(x,y)$ and $\Delta w(x,y)$, the material parameters $t(x,y)$ and $\langle n_g(x,y)\rangle_z$ can be obtained simultaneously as

$$t(x,y) = \frac{\Delta D(x,y) - \Delta w(x,y)}{n_{air}}, \tag{14}$$

$$\langle n_g(x,y)\rangle_z = \frac{\Delta D(x,y)}{\Delta D(x,y) - \Delta w(x,y)} \cdot n_{air}. \tag{15}$$

$\langle n_g(x,y)\rangle_z$ and $t(x,y)$ depict the lateral refractive index distribution and physical thickness uniformity of a film sample.

4.2 Simultaneous film thickness and refractive index metrology by hybrid confocal-scan FD-OCT

By incorporating confocal-scanning-based focus tracking into an FD-OCT system, the refractive index and geometrical thickness of a film layer can be simultaneously estimated. This method is applicable to measure the refractive index and thickness of each individual layer within multi-layer film samples.

To exemplify the method, measurements on a single film sample are illustrated in **Figure 7**. By measuring the distance between the peaks of the two intensity PSFs on an FD-OCT depth profile as shown in **Figure 7b**, the group optical thickness ΔD between the top and bottom surfaces of the layer can be obtained, which is expressed as

$$\Delta D = n_g \cdot t, \tag{16}$$

where n_g is the group refractive index, and t is the physical layer thickness.

In order to extract simultaneously the group index n_g and the physical layer thickness t, an additional independent relationship between these two quantities is required. In a hybrid confocal-scan FD-OCT system, the sample arm is configured as shown in **Figure 7a**. This setup allows to obtain the additional information of the focal shift distance (Δz) needed to focus the objective lens from the top surface of the layer to the bottom surface, which yields an additional relationship between, in this case, the phase index of refraction n_p and the thickness t given as

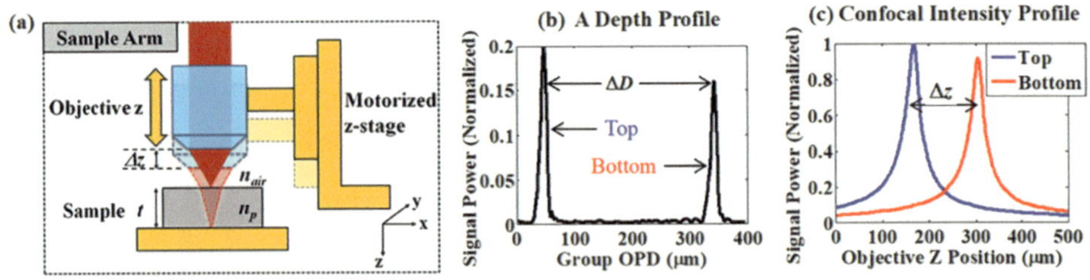

Figure 7.
(a) Illustration of the sample arm of a hybrid confocal-scan FD-OCT system during measurement through a single film. (b) An example of an FD-OCT depth profile acquired when the objective is focused between the top and bottom surfaces. (c) Confocal intensity profiles of both the top and bottom surfaces reconstructed from a sequence of depth profiles acquired simultaneously with the objective lens being translated to focus through the top and bottom surfaces (adapted from [26]).

$$n_p = \sqrt{NA^2 + \left(n_{air}^2 - NA^2\right)\left(\frac{t}{\Delta z}\right)^2}, \tag{17}$$

where NA is the numerical aperture of the objective lens, and n_{air} is the phase index of the ambient air. In practice, Δz is obtained from the confocal intensity profiles of both the top and bottom surfaces as shown in **Figure 7c**, which are reconstructed in post-processing by tracing the peak amplitude of the respective PSFs on a sequence of depth profiles acquired, while the objective lens is translated to focus through the top and bottom surfaces.

Note that the refractive index involved in the confocal-scan Δz measurement is the phase index, whereas the FD-OCT ΔD measurement concerns the group index. To obtain independently the phase index, group index, and thickness of a layer, the conversion between the group and phase indices is required as described by the dispersion relation

$$n_g = n_p - \Delta n_{disp} = n_p - \lambda \frac{\partial n_p}{\partial \lambda}. \tag{18}$$

For common polymeric films, the material dispersion property, that is, $\frac{\partial n_p}{\partial \lambda}$, is assumed to be a priori knowledge that can be directly applied in Eq. (18). If the dispersion term is unknown, it can be obtained by conducting the confocal-scan FD-OCT measurements separately at another spectral band (e.g., around both 840 nm and 1300 nm). The same principle works for additional spectral bands to measure the index, thickness, and dispersion characteristic of unknown samples.

Combining Eqs. (16)–(18), n_p can be computed by solving the following quartic equation:

$$\begin{gathered} An_p^4 + Bn_p^3 + Cn_p^2 + Dn_p + E = 0, \\ A = (\Delta z)^2, \\ B = -2\Delta n_{disp}(\Delta z)^2, \\ C = \left[\left(\Delta n_{disp}\right)^2 - (NA)^2\right](\Delta z)^2, \\ D = 2\Delta n_{disp}(\Delta z)^2(NA)^2, \\ E = -\left(\Delta n_{disp}\right)^2(\Delta z)^2(NA)^2 - \left[n_{air}^2 - (NA)^2\right](\Delta D)^2. \end{gathered} \tag{19}$$

Once n_p is computed, n_g and t can be solved consecutively based on Eqs. (18) and (16).

5. Examples

5.1 3-D volumetric imaging of films

Ultra-high-resolution OCT systems, exemplified by the Gabor domain optical coherence microscopy (GD-OCM) [28, 29], enable nondestructive volumetric inspection of fine structures within otherwise transparent-appearing film samples [30].

The GD-OCM system utilizes a liquid lens embedded in a custom optical system designed to achieve dynamic focusing of the component at both axial and lateral optical resolutions of 2 μm throughout up to 2 mm. The layout of the system is detailed elsewhere [28]. The light source is a Titanium:Sapphire femtosecond laser centered at 800 nm with 120 nm FWHM (Integral, Femtolasers Inc.). The spectral interference signal is registered onto up to 4992 pixels (10 μm × 20 μm per pixel) of a high-speed CMOS line camera (spl8192-70 km, Basler Inc.) through a custom spectrometer. The exposure time was set to 20 μs and the acquisition speed was 23,000 A-lines/sec with x-y scanning by two cross-axis mirrors driven by two galvanometers synchronized to the acquisition of the CMOS camera.

In this example, two films under investigation are 50-μm thick SAN17/PMMA co-extruded polymer films, bounded by 25-μm thick protective layers on both sides. Both films, labeled as film #1 and film #2, are of the same composition, yet processed under different temperatures and pressures during fabrication. With 2-μm resolution, miniscule line and particle defects are clearly identified inside the films by the GD-OCM system. Furthermore, the locations and dimensions of the defects in 3-D are precisely measured.

Figure 8a and **b** show 3-D GD-OCM imaging of two films rendered by Voxx software (The School of Medicine, Indiana University). Film #1 exhibits more densely distributed defects compared to film #2, as shown by an ocean of particulates across the entire film volume. **Figure 8c** and **d** are two examples of enlarged *en face* (*x-y* plane) and corresponding cross-sectional (*x-z* plane) images of a few typical defects of film #2 as denoted by red arrows in **Figure 8b**. The lateral extent of the defects can be identified from the *en face* images, whereas their depth locations and extent can be accurately quantified from the cross-sectional images.

A discovery based on this metrology was that, for otherwise equivalent films, the processing temperature and pressure applied during manufacture have an important impact on the defect level in the films.

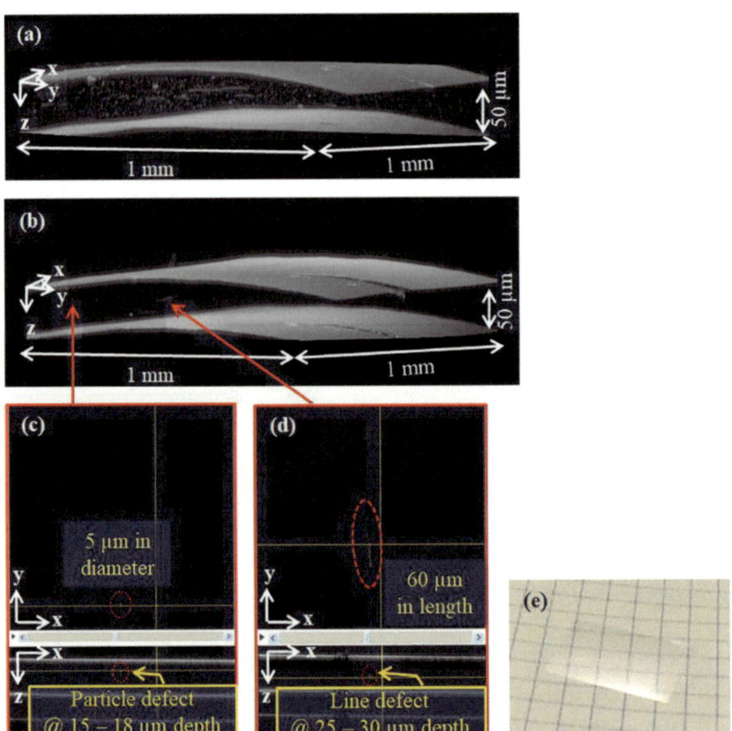

Figure 8.
GD-OCM imaging of two identical composition 50-μm thick films extruded under different temperatures and pressures. (a) Film #1 showing a multitude of defects. (b) Film #2 showing few defects. (c) and (d) are two examples of enlarged en face *and corresponding cross-sectional images of a few typical defects of film #2 in the locations denoted by red arrows in (b). (e) Photograph of a film (adapted from [30]).*

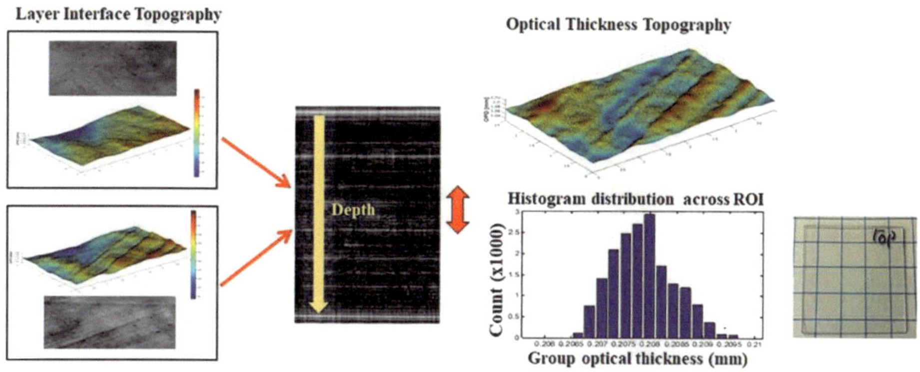

Figure 9.
Illustration of film layer interfaces and optical thickness topography with statistical distribution denoted.

5.2 Film interface and thickness topography

From 3-D volumetric imaging data of film samples, accurate topography of the surfaces of a single-layer film or interfaces among a multi-layer film sample can be readily extracted by various peak detection or surface segmentation algorithms. Such topographies allow nondestructive inspection of the spatial uniformity and surface waviness of samples, which carries important value for pinpointing the locations and root causes of surface defects, particularly useful within multi-layer samples.

Figure 9 shows an example of profiling the inter-layer interfaces within a multi-layer film stack sample that visually appears transparent. The data were acquired from an SS-OCT system with an axial PSF of approximately 8 μm FWHM in air and a lateral resolution of 20 μm. The NIR light source (Micron Optics, Inc.) periodically sweeps over a wavelength range of 1240–1400 nm at a constant sweep rate of 45 kHz. The instantaneous linewidth of the source is about 0.2 nm, resulting in an imaging depth range of about 1 mm. Based on the acquired OCT imaging data, a surface segmentation algorithm is applied to produce 3-D visualization of the surface profiles of inter-layer interfaces. The difference in the topography of two interfaces leads to the optical thickness topography of film layer stack.

5.3 Metrology of film thickness through depth

As has been shown, the superb depth sectioning capability of OCT reveals not only the surfaces but also the layer interfaces within a film sample where specular or scattering reflection signals are generated from either layer-to-layer refractive index change or lamination imperfection.

A $33 \times 33 \times 2.93$ mm^3 (x, y, z) multi-layered monolithic sheet sample was examined by an SS-OCT system with ~1 mm depth range as described in Section 5.2 [31]. The sample is composed of 108-layer polymer films of coextruded 10%/90% PMMA/SAN17 with nominal individual film thickness of 27 μm. Two sets of 3-D OCT data covering the top and bottom portions of the sample were collected and volumetrically rendered as shown in **Figure 10**. From the 3-D OCT data sets, a lateral location was selected where the group optical thickness of each film layer imaged by the SS-OCT system is computed and divided by the group refractive index of the sample to yield estimated geometrical thickness. **Figure 10f** and **g** are two plots of the layer thickness profiles as a function of the layer number counted from the top and the bottom surfaces of the sample, respectively, which indicate

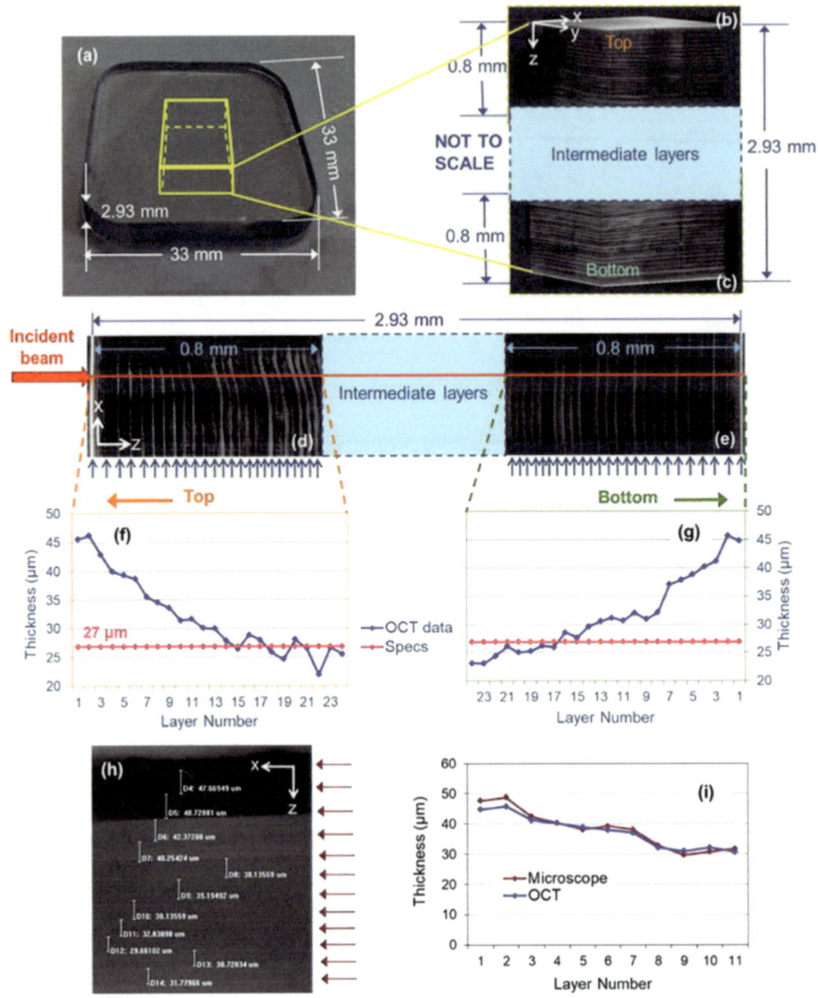

Figure 10.
Nondestructive metrology of the layer thickness profiles over the depth of a multi-layer monolithic polymeric sample. (a) A photograph of the sample. (b) and (c) are volumetric rendering of the 3-D OCT data sets of the top and bottom portions of the sample, respectively. (d) and (e) are cross-sectional OCT images of the top and bottom portions of the sample with the red line representing the location where OCT depth scans were taken for thickness estimation. (f) and (g) are layer thickness profiles of the top and bottom 23 layers, respectively. (h) A cross-sectional image of the bottom 11 layers of the sample after being cut and imaged under a light microscope. (i) Quantitative comparison of the layer thickness profiles of the bottom 11 layers obtained from OCT and microscope measurements (adapted from [31]).

that the layer thicknesses decrease from the near surface layers toward the inner layers of the sample.

The thickness measurements were confirmed by cutting the sample, polishing the cut surface, and imaging it under a light microscope with 20x magnification as shown in **Figure 10h**. From the dense layer structure imaged by the microscope, the thicknesses of the first 11 layers counting from the bottom surface were measured. The film thicknesses measured from the microscope image and those measured by SS-OCT over the same region of the sample are plotted in **Figure 10i**, which shows good agreement. The validation proves OCT as a compelling technique for nondestructive characterization of the layer thickness distribution in multi-layered polymeric material sheets despite their apparent full visual transparency.

5.4 Simultaneous refractive index and thickness metrology

Conventional OCT systems rely on a priori knowledge of the refractive index of the sample under test to estimate its geometrical thickness. Hybrid confocal-scan SS-OCT systems may provide simultaneous measurements of both the refractive index and thickness of a single- or multi-layer film sample as described in Section 4.2.

Figure 11 shows an example of the simultaneously measured refractive index and layer thickness profiles of a sample consisting of 108-layer coextruded 10%/ 90% PMMA/SAN17 films as described in Section 5.3. The OCT system incorporates a confocal scanning mechanism with 0.1-μm z-resolution in an SS-OCT system with ~5 mm depth range. The swept laser (HSL-2100-WR, Santec, Japan) is centered at 1318 nm with a FWHM spectral bandwidth of 125 nm and an instantaneous linewidth of 0.1 nm. The axial PSF of the system is approximately 10 μm FWHM in air and the effective frequency sweep rate of the source is 20 kHz.

During the confocal scan, each film surface or layer interface under test went in and out of focus, consecutively. Confocal intensity profiles were reconstructed from the entire sets of depth profiles collected to show the back-reflection intensity variations for all interfaces as a function of the axial position of the objective lens. Combining the measurements of the group optical thickness ΔD between two interfaces provided by the depth profiles and the translation Δz needed to focus from one interface to the next, the refractive index and thickness between the two interfaces were simultaneously computed. To validate the measurement consistency and uncertainty through increasing the thickness, the first surface was fixed as the top surface of the monolithic film-stack sample and the second surface was varied along depth from the first layer interface successively to the bottom surface of the sample. The measured phase refractive index and the cumulative thickness of

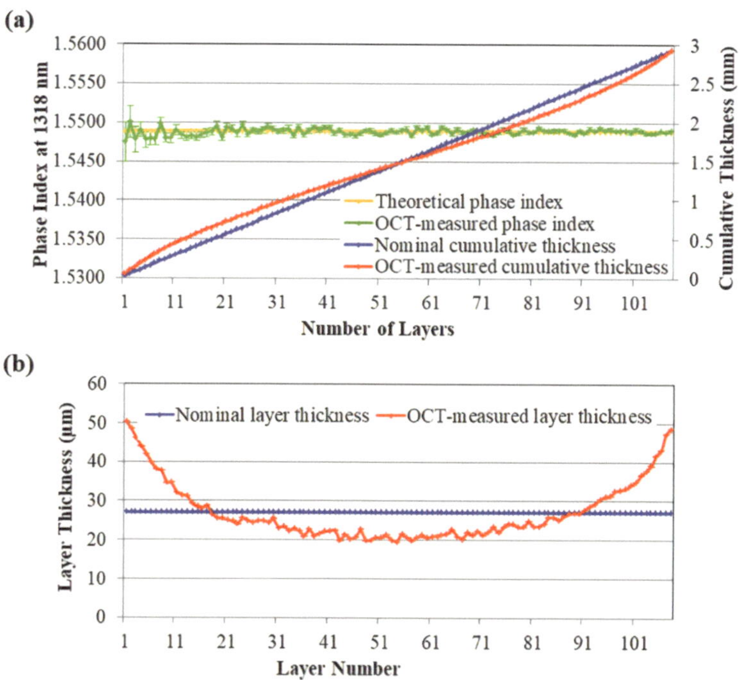

Figure 11.
(a) Theoretical and OCT-measured phase refractive index and cumulative thickness of a monolithic 108-layer 10/90% PMMA/SAN17 sample measured across increasing numbers of layers. (b) Nominal and OCT-measured thickness of each layer (adapted from [26]).

1–108 layers averaged from 100 repeated measurements are plotted in **Figure 11a**, together with their standard deviations, against the theoretical values. Based on the composition of the monolith, it is predicted to have a phase index of 1.5489, while the OCT-measured index (average of 100 measurements) ranges from 1.5475 to 1.5500 and deviates from the theoretical value by about 0.00003 (measuring 106 layers, 2.8378 mm thick) to 0.0014 (measuring 1 layer, 0.0502 mm thick). The standard deviation of 100 measurements varies from 0.0001 (measuring 105 layers, 2.7946 mm thick) to 0.0025 (measuring 1 layer). It is also shown in **Figure 11a** that the measured cumulative thickness increases more rapidly near both surfaces of the sample. The standard deviation of the measured cumulative thicknesses across 100 measurements varies from 0.12 μm (measuring one layer) to 0.28 μm (measuring 105 layers).

The estimated thickness of each individual film layer was obtained by computing the difference between the cumulative thicknesses of two consecutive numbers of layers and is plotted in **Figure 11b**, which shows a nearly parabolic profile as opposed to a theoretical uniform layer thickness of ~27 μm. The non-uniformity in the measured film layer thicknesses helped to diagnose the issue in the fabrication process and facilitated the improvement of the process.

6. Summary and perspectives

Development of functional polymer films and film stacks has been under increasing demand to create new generations of novel, compact, light-weight optics. OCT provides the right tool for the metrology of all the key optical properties of these films, ranging from flat to curved geometries. In this chapter, the system design, metrology methodologies, and examples of OCT for film metrology are discussed to provide both the knowledge foundation and the engineering perspectives. The advanced film metrology capabilities offered by OCT, such as simultaneous refractive index and thickness estimation, and 3-D inspection of curved films by angular-scan OCT [14, 15], play a key role in the manufacturing process maturity of newly developed films. Rapid advancement in the field of OCT is foreseen to drive the application toward in-line film metrology and facilitate the rapid growth of innovative films in the industry.

Author details

Jianing Yao* and Jannick P. Rolland
The Institute of Optics, University of Rochester, Rochester, NY, USA

*Address all correspondence to: jianingyao.optics@gmail.com

References

[1] Häusler G, Lindner MW. "Coherence Radar" and "Spectral Radar"—New tools for dermatological diagnosis. Journal of Biomedical Optics. 1998;**3**(1): 21-31

[2] Wojtkowski M, Leitgeb R, Kowalczyk A, Bajraszewski T, Fercher AF. In vivo human retinal imaging by Fourier domain optical coherence tomography. Journal of Biomedical Optics. 2002;**7**(3):457

[3] Nassif NA, Cense B, Park BH, Pierce MC, Yun SH, Bouma BE, et al. In vivo high-resolution video-rate spectral-domain optical coherence tomography of the human retina and optic nerve. Optics Express. 2004;**12**(3):367-376

[4] Chinn SR, Swanson EA, Fujimoto JG. Optical coherence tomography using a frequency-tunable optical source. Optics Letters. 1997;**22**(5):340-342

[5] Golubovic B, Bouma BE, Tearney GJ, Fujimoto JG. Optical frequency-domain reflectometry using rapid wavelength tuning of a Cr4+:forsterite laser. Optics Letters. 1997;**22**(22):1704-1706

[6] Yun SH, Tearney GJ, de Boer JF, Iftimia N, Bouma BE. High-speed optical frequency-domain imaging. Optics Express. 2003;**11**(22):2953-2963

[7] Huber R, Wojtkowski M, Fujimoto JG. Fourier domain mode locking (FDML): A new laser operating regime and applications for optical coherence tomography. Optics Express. 2006; **14**(8):3225-3237

[8] Tearney GJ, Bouma BE, Fujimoto JG. High-speed phase- and group-delay scanning with a grating-based phase control delay line. Optics Letters. 1997; **22**(23):1811-1813

[9] Iyer S, Coen S, Vanholsbeeck F. Dual-fiber stretcher as a tunable dispersion compensator for an all-fiber optical coherence tomography system. Optics Letters. 2009;**34**(19):2903-2905

[10] Fercher AF, Hitzenberger CK, Sticker M, Zawadzki R, Karamata B, Lasser T. Dispersion compensation for optical coherence tomography depth-scan signals by a numerical technique. Optics Communications. 2002;**204** (1–6):67-74

[11] Marks DL, Oldenburg AL, Reynolds JJ, Boppart SA. Digital algorithm for dispersion correction in optical coherence tomography for homogeneous and stratified media. Applied Optics. 2003;**42**(2):204-217

[12] Wojtkowski M, Srinivasan VJ, Ko TH, Fujimoto JG, Kowalczyk A, Duker JS. Ultrahigh-resolution, high-speed, Fourier domain optical coherence tomography and methods for dispersion compensation. Optics Express. 2004; **12**(11):2404-2422

[13] Qi B, Himmer AP, Gordon LM, Yang XV, Dickensheets LD, Vitkin IA. Dynamic focus control in high-speed optical coherence tomography based on a microelectromechanical mirror. Optics Communications. 2004;**232** (1–6):123-128

[14] Yao J, Meemon P, Ponting M, Rolland JP. Angular scan optical coherence tomography imaging and metrology of spherical gradient refractive index preforms. Optics Express. 2015;**23**(5):6428-6443

[15] Yao J, Thompson KP, Ma B, Ponting M, Rolland JP. Volumetric rendering and metrology of spherical gradient refractive index lens imaged by angular scan optical coherence tomography system. Optics Express. 2016;**24**(17): 19388-19404

[16] Wang D, Liang P, Samuelson S, Jia H, Ma J, Xie H. Correction of image

distortions in endoscopic optical coherence tomography based on two-axis scanning MEMS mirrors. Biomedical Optics Express. 2013;**4**(10): 2066-2077

[17] Ortiz S, Siedlecki D, Remon L, Marcos S. Optical coherence tomography for quantitative surface topography. Applied Optics. 2009;**48** (35):6708-6715

[18] Malacara D. Optical Shop Testing. 3rd ed. Hoboken, NJ: Wiley; 2007

[19] Yao J, Anderson A, Rolland JP. Point-cloud noncontact metrology of freeform optical surfaces. Optics Express. 2018;**26**(8):10242-10265

[20] Tearney GJ, Brezinski ME, Southern JF, Bouma BE, Hee MR, Fujimoto JG. Determination of the refractive index of highly scattering human tissue by optical coherence tomography. Optics Letters. 1995;**20**(21):2258-2260

[21] Uhlhorn SR, Borja D, Manns F, Parel J-M. Refractive index measurement of the isolated crystalline lens using optical coherence tomography. Vision Research. 2008;**48**(27):2732-2738

[22] Verma Y, Rao KD, Suresh MK, Patel HS, Gupta PK. Measurement of gradient refractive index profile of crystalline lens of fisheye in vivo using optical coherence tomography. Applied Physics B: Lasers and Optics. 2007;**87**(4):607-610

[23] de Castro A, Ortiz S, Gambra E, Siedlecki D, Marcos S. Three-dimensional reconstruction of the crystalline lens gradient index distribution from OCT imaging. Optics Express. 2010;**18**(21):21905

[24] Ohmi M, Ohnishi Y, Yoden K, Haruna M. In vitro simultaneous measurement of refractive index and thickness of biological tissue by the low coherence interferometry. IEEE Transactions on Biomedical Engineering. 2000;**47**(9):1266-1270

[25] Kim S, Na J, Kim MJ, Lee BH. Simultaneous measurement of refractive index and thickness by combining low-coherence interferometry and confocal optics. Optics Express. 2008;**16**(8):5516

[26] Yao J, Huang J, Meemon P, Ponting M, Rolland JP. Simultaneous estimation of thickness and refractive index of layered gradient refractive index optics using a hybrid confocal-scan swept-source optical coherence tomography system. Optics Express. 2015;**23**(23): 30149-30164

[27] Zhou Y, Chan KK, Lai T, Tang S. Characterizing refractive index and thickness of biological tissues using combined multiphoton microscopy and optical coherence tomography. Biomedical Optics Express. 2013;**4**(1): 38-50

[28] Lee KS, Thompson KP, Meemon P, Rolland JP. Cellular resolution optical coherence microscopy with high acquisition speed for in-vivo human skin volumetric imaging. Optics Letters. 2011;**36**(12):2221-2223

[29] Canavesi C, Rolland JP. Ten years of Gabor-domain optical coherence microscopy. Applied Sciences. 2565;**9** (12):2019

[30] Meemon P, Yao J, Lee KS, Thompson KP, Ponting M, Baer E, et al. Optical coherence tomography enabling non-destructive metrology of layered polymeric GRIN material. Scientific Reports. 2013;**3**:1709

[31] Yao J, Meemon P, Lee KS, Rolland JP. Nondestructive metrology by optical coherence tomography empowering manufacturing iterations of layered polymeric optical materials. Optical Engineering. 2013;**52**(11):112111

Chapter 6

Fouling Monitoring in Membrane Filtration Systems

Luca Fortunato

Abstract

Membrane filtration systems are employed in the water industry to produce drinking water and for advanced wastewater treatment. Fouling is considered the main problem in membrane filtration systems. Fouling occurs when the biomass deposited on the membrane surface leads to a membrane performance decline. Most of the available techniques for characterization of fouling involve the analysis of membrane samples after membrane autopsies. This approach provides information ex-situ destructively at the end of the filtration process. Optical coherence tomography (OCT) gained attention in the last years as noninvasive imaging technique, capable of acquiring scans in-situ and nondestructively. The online OCT monitoring enables visualizing and studying the biomass deposition over time under continuous operation. This approach allows to relate the impact of the fouling on the process. In the last years, the suitability of OCT as in-situ and nondestructive tool for the study of fouling in membrane filtration systems has been evaluated. The OCT has been employed to study the fouling in different membrane geometry and configuration for the treatment of seawater and wastewater. Nowadays, the OCT is employed to better understand the role of biomass structure on the filtration mechanisms.

Keywords: OCT, water treatment, desalination, biofouling, fouling, membrane filtration

1. Water scarcity

Nowadays, the insufficiency of access to clean and secure water represents one of the main problems for sustainable development affecting both industrialized and developing countries. It is estimated that by 2030, almost 50% of the global population will face water stress conditions. Although 71% of the planet Earth's surface is covered by water, only less than 3% is constituted by fresh water, where the remaining 97% of the water is seawater and characterized by the high content of mineral salt, which makes it inadequate for direct consumption.

The continuous increase in the human population and industrialization lead to a constant increase of the water demand. Freshwater is not used only for direct consumption, but it is the central pillar of food and energy production. The use of water in agriculture contributes to 70% of the total water withdrawn [1]; the other major contribution to the global consumption is related to the energy production, whereas water is required in the whole cycle of the energy production and distribution. Besides the scarcity, another major threat is represented by the quality of the water available. According to the UNESCO, almost 3.6 billion people, lack for access to clean water and proper sanitation [2]. To meet the continuous demand, there is a need to produce freshwater starting by nonconventional sources, as saline water (seawater and brackish) and other contaminated fresh water sources (wastewater and industrial water). Among the different technologies tested over the years, membrane filtration is recognized as the most common and convenient method to purify water.

1.1 Membrane technology

The membrane is a semi-permeable barrier used to separate specific substances considered as pollutants from the water. The purification occurs by applying pressure in the systems, whereas the membrane allows only certain ions and molecules to be transported through the membrane with a specific size to pass, retaining all the rest. During the filtration, the water passes through the membrane and the rejection of other compounds depends on the pressure applied and the membrane pore size. The main membrane filtration processes and their respective removal capabilities are shown in **Figure 1**. By using different membranes, it is possible to remove specific unwanted compounds. Hence, safe drinking water can be produced starting from different sources, including seawater and wastewater.

Membranes can be classified according to different criteria. The most diffused classification is based on the membrane pore size, whereas the removal of unwanted contaminants is related to the membrane pore size. The highest removal is realized by reverse osmosis membranes (RO) followed by nanofiltration membrane (NF), are generally used to remove salt from water (desalination from seawater). Ultrafiltration (UF) and microfiltration (MF) membranes, with pore size from 0.001 to 0.1 μm, are employed to remove pathogens and suspended solids. RO membrane requires high pressure up to 70 bar to remove the salt during desalination processes, while MF and UF usually require lower pressure less than 5 bar [3].

Another classification is based on the geometry of the membrane employed in the process (i.e., flat sheet, spiral wound, tubular, and hollow fiber). The use and choice of a specific geometry are often linked to a specific application and system design,

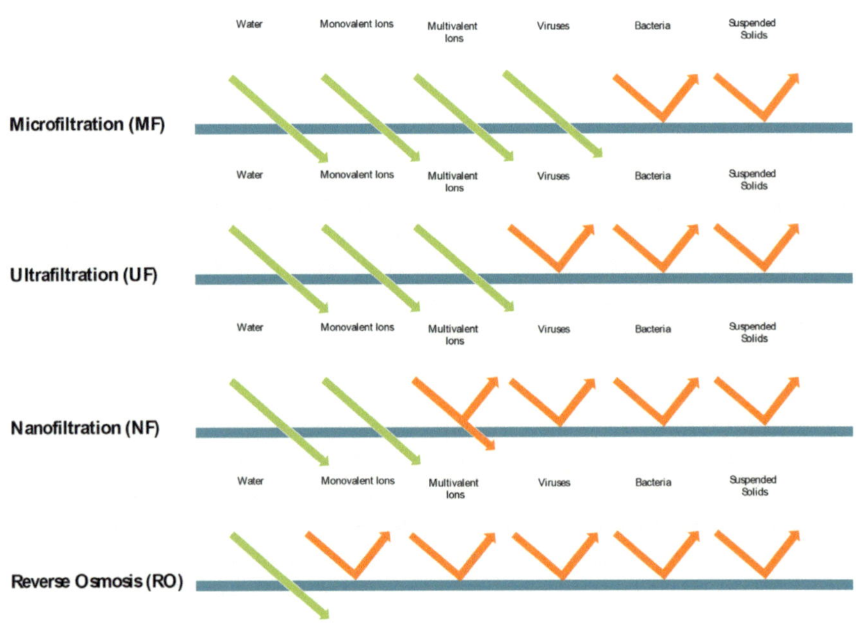

Figure 1.
Membrane size and compounds separation.

where the membrane module can be submerged or external. The most employed submerged membrane configuration is the membrane bioreactor (MBR), whereas the membrane unit can be directly submerged in the aeration tank or in a separate tank. On the contrary, commercial seawater membrane desalination plants employ RO membrane as external module inserted in vessels that allows the use of high pressure (60–70 bar) necessary to remove salt from water. These membranes are called spiral wound membrane modules, and consist of membrane sheets enfolded along a central tube, comprising a perforated central tube for permeate collection surrounded by layers of membrane, permeate spacers and feed spacers. The produced water, called permeate, is collected in the central tube by the product spacer.

1.2 Membrane fouling

The continuous filtration of water over time leads to the accumulation of rejected material on the membrane surface and or in membrane pores. This phenomenon is called fouling. The development of membrane fouling on the membrane surface is considered the bottleneck of membrane filtration processes. Over time, fouling is inevitable and leads to the decrease of the membrane flux that is considered as the main process performance indicator. The decrease in flux is followed by the energy increase as the pressure applied to overcome the reduction in water production. Moreover, to recover the membrane permeability, the operators increase the use of chemicals for the cleaning, therefore reducing the membrane lifetime. Therefore, the fouling understanding and control are considered the major challenges encountered in membrane filtration processes as highlighted by the crescent number of scientific publications and the increase of commercial products for the reduction and control of fouling.

The fouling deposited in membrane filtration systems varies depending on each specific process, and it is due to the complex interaction between the constituents present in the feed water and the membrane [4, 5]. The fouling mechanism generally involves (i) initial pore blocking followed by (ii) cake layer formation. Depending on the characteristics of the water treated, the fouling in membrane filtration systems can be divided into four categories (**Figure 2**):

- Particulate fouling
- Organic fouling
- Inorganic fouling (or scaling)
- Biofouling

Particulate fouling is due to the accumulation of suspended particles of different nature on the membrane surface. Organic fouling consists of the formation of an organic layer on the surface and is linked to the concentration of dissolved organic

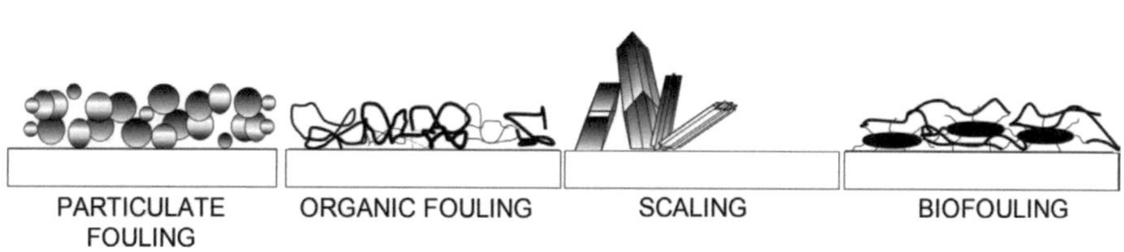

Figure 2.
Different types of fouling deposited on the membrane surface.

compounds in the feed water. The scaling is the most common type of inorganic fouling in desalination processes; it is due by the direct formation of crystals on the membrane surface through precipitation. Depending on the process, different types of fouling can occur concurrently; in that case, often the fouling deposited on the membrane is referred as biomass.

Among the different types of fouling membrane, biofouling is considered the most problematic fouling faced in membrane filtration system, which negatively affects the process in terms of technology and economics [6]. Biofouling refers to the development of biofilms in the membrane systems and it is caused by the accumulation of microorganisms, including extracellular polymeric substances (EPS) produced by microorganisms, on a surface due to either deposition and/or growth. Biofouling is defined to occur when the biofilm passes a threshold of interference negatively affecting the filtration process [7]. The development and growth occur due to the continuous availability of nutrients flowing into the system [8].

Biofouling is considered the least understood and most problematic type of fouling in affecting filtration processes despite the crescent of a significant number of studies [9].

2. Fouling characterization

Over time, the formation of fouling on the membrane surface acts as a secondary filtration layer impacting the membrane performance due to the increase of the hydraulic resistance of the system. During operation, the only information available regarding the fouling formation is represented by the decrease in performance, either a decrease in flux or an increase in pressure, without any data related to the identity of the fouling (**Figure 3**). Hence, either on a full-scale plant or in a research experiment, it is common to conduct membrane autopsy in order to analyze membrane coupons collected from the module. This approach involves two different types of analysis, (i) analytical characterization performed to identify the nature of the contaminants and (ii) visualization by means of an imaging technique to identify and quantify the fouling structure. A key aspect of fouling studies includes the analysis of the structural fouling properties [10], which can forecast the fouling layer comportment, and thus, the effect on filtration performance. The structural analysis of the fouling layer deposited on a membrane coupon consists of using imaging techniques. Most of the imaging approaches and practices to characterize the fouling morphology reported in the literature include require destructive procedures and are performed after membrane autopsy (**Figure 4**) [11, 12].

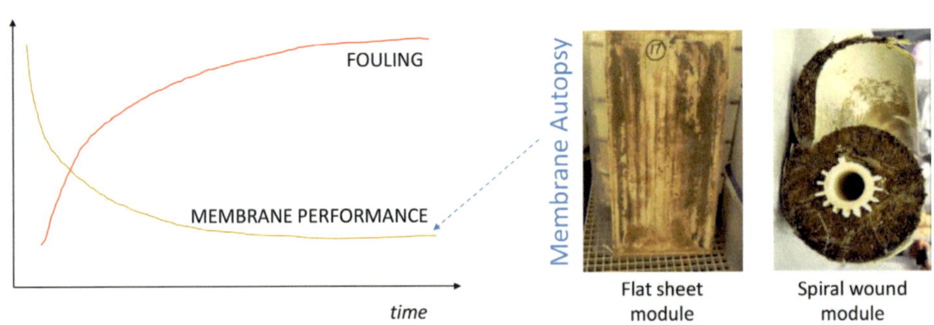

Figure 3.
Fouling development and membrane performance decrease in a membrane module.

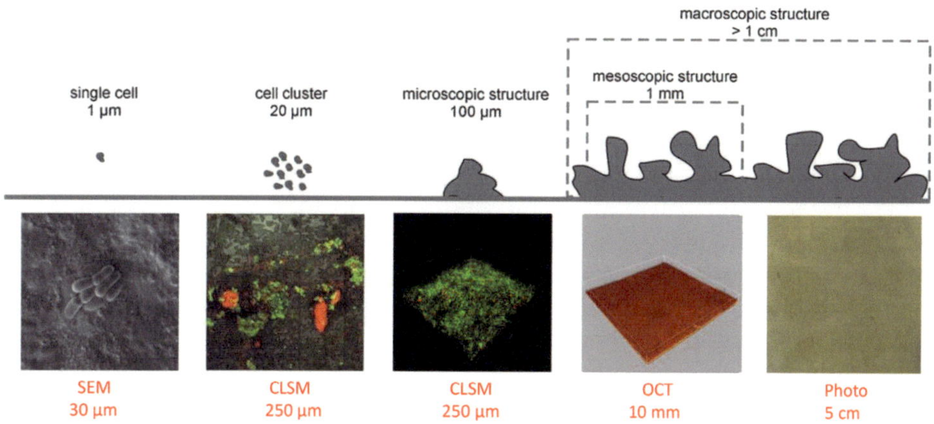

Figure 4.
Techniques to characterize fouling structure at different scales: SEM, scanning electron microscopy; CLSM, confocal laser scanning microscopy; and OCT, optical coherence tomography. Adapted from Wagner et al. [18].

Scanning electron microscopy (SEM) has been for many years the most employed technique to characterize membrane coupons. SEM is capable of visualizing the structure of fouling on a near-nanometer and sub-micron scale, allowing to distinguish among the different types of fouling, including crystals and biofilms [13, 14]. However, the SEM analysis requires the drying and coating of the analyzed sample, which represents a limitation in terms of time and artifacts. Confocal laser scanning microscopy (CLSM) has been widely used for the analysis of biological samples (biofouling). In particular, CLSM enables characterizing the constituents of biofilm, including the EPS matrix with the use of specific probes [15]. Compared to SEM, CLSM allows the three-dimensional (3D) characterization of the fouling layer [16]. However, the sample preparation for the CLSM is more complex requiring the use of specific dyes and probe, with the risk of altering and affecting the overall structure. As highlighted in the literature, all the steps involved in the staining, including storage and rinsing, can modify the biofilm structure and thus impact parameter quantification [17].

Moreover, the CLSM allows the quantification and visualization only of the stained material that is only partially representative of the fouling structure. Therefore, considering the sample handling and the staining process, the structural analysis performed by following this approach is not truly representative of fouling deposited in the system. An additional limitation is represented by the incapacity of providing information online during continuous operation.

2.1 In-situ nondestructive fouling characterization

The main limitation of the conventional imaging techniques in characterizing the fouling is represented by the impossibility of collecting data over time. In fact, SEM and CLSM are destructive analysis, requiring the destruction of the operating modules. Therefore, the destructive approach enables collecting data "only once" by destroying the membrane (**Figure 3**). Usually, the analysis is performed by membrane autopsy at the end after a significant decrease in performance, consisting in most of the cases in a decrease in permeability, that is translated in a decrease in flux under constant pressure operation and an increase in pressure under constant flux operation. Therefore, as shown in **Figure 3**, the membrane autopsy is performed usually at the end of the experiment or toward the minimum of its performances, when is "too late." In summary, conventional techniques lack in providing temporal information related to the fouling developed in the system.

Therefore, in membrane filtration processes, nondestructive in-situ biofilm fouling monitoring techniques have gained attention thanks the possibility of obtaining information regarding the fouling formed in membrane systems without stopping the process or destroying the units. Several techniques have been tested over time. The first approach consisted of employing a camera mounted on the system to monitor the membrane surface. This method allowed to evaluate the nondestructive fouling deposition and pattern at particular operating conditions. Another approach employed nuclear magnetic resonance (NMR) to detect the morphological development of biomass in membrane system [19, 20]. Recently, another approach was proposed employing planar optodes to visualize the biomass in a membrane flow cell under continuous operation by using probes. This approach enables to assess the O_2 distribution in biofilms and therefore estimate the biofouling spatial distribution [21]. Recently, the OCT has been tested as tool to evaluate the fouling deposition in membrane filtration systems in-situ nondestructively.

3. Biofilm characterization with OCT

The OCT was first developed and mainly employed in biomedical applications. The technology has been extensively used in ophthalmology for diagnosis and treatment guidance [22]. Afterward, several applications were explored, including the study of biofilm structure. In 2006, the OCT was utilized to study the growth of a *Pseudomonas aeruginosa* biofilm in a capillary flow cell [23]. Haisch and Niessner [24] evaluated the suitability of the OCT for industrial biofilm monitoring. Wagner et al. [18] showed the potentiality of a SD-OCT system OCT in characterizing the biofilm structure in the millimeters range. As highlighted in Section 1.2, the formation of biofilm is very common in membrane processes; therefore, these studies laid the foundation for studying the biomass growth in membrane processes.

Over time, the use of the OCT gained significative attention in the study of biofouling due to a series of advantages respect the other conventional techniques [11, 25]. The first benefit is due to the ability to investigate fouling formation in-situ without any staining. Indeed, the OCT is a label-free technique that enables acquiring 3D data eliminating the all risks related to the use of specific probe and chemicals [26, 27]. Another advantage consists in the possibility of acquiring information at the mesoscale level, allowing to monitor the biomass in the millimeters range. The mesoscale is considered ideal for studying the bulk-fouling interface in order

Figure 5.
3D rendered fouling structure obtained from the OCT and CLSM datasets. Adapted from Fortunato et al. [28].

to understand the fluid-structure interaction (**Figure 3**). On the other side, SEM and CLSM allow to acquire information at the microscopic level in the micrometers range. The diminution of the monitored portion increases the risk of analyzing an area not representative of the process. By using CLSM, we need to acquire more than 400 scans to cover the same area covered by a single OCT scan (**Figure 5**) [18, 28]. Moreover, it is worth to mention that the OCT does not require any time for sampling and a 3D scan can be acquired in the range of 1–2 minutes. All these features are necessary to acquire information regarding the fouling formation to assess the amount of biomass deposited in the system and its hydraulic resistance of the layer deposited on the membrane. Indeed, the main objective of the fouling characterization is to understand the impact of the biomass deposited on the system performance [10].

3.1 Morphology analysis

Besides the visualization, the main objective of the OCT in-situ observation is related to the possibility of describing and quantifying the monitored structure. A complete list of the key parameters that can be obtained from tomography dataset was previously presented by Beyenal et al. [29]. The OCT device used is in most of the studies is the Thorlabs GANYMEDE spectral domain OCT system with a central wavelength of 930 (Thorlabs, GmbH, Dachau, Germany) equipped with a 5× telecentric scan lens (Thorlabs LSM 03BB). In the case of fouling characterization, the most used parameters are: mean thickness, relative roughness, absolute roughness, membrane coverage, biovolume, and macro-porosity. The relative and absolute roughnesses are parameters used to assess the heterogeneity of the biomass morphology. The first step of the image analysis necessary to extract the data is binarization, which allows to identify and distinguish the biomass signal to the background noise. This step in the case of OCT datasets is complex since in most of the cases it is not possible to perform the segmentation based only on a pixel intensity threshold. Indeed, the polymeric support below the membrane has often the same intensity of the biomass deposited on the membrane surface. Therefore, often the intensity threshold needs to be coupled with an edge detection algorithm that allows to identify biomass, only the pixels above the membrane. Another obstacle is due to the speckle noise that hinders the biomass binarization. Frequently *mean* and *median* filters are applied before the thresholding to improve the binarization processes. However, these filters are not suitable for interferometric images and might alter the OCT dataset. Several efforts have been made to develop algorithms to denoise the OCT scans [30]. Recently, a specific algorithm was developed and successfully tested for the study and the quantification of the cake layer development in activated sludge membrane bioreactor treating real wastewater [28, 31].

An alternative strategy to simplify the binarization process and the image analysis consists in subtracting the initial OCT scan acquired at time 0 to all the other scans, in this way all the extern signals to the biomass are eliminated from the scans. This approach has been successfully employed in different cases and membrane processes [32–35].

The first parameters usually calculated from the OCT datasets is the fouling mean thickness. This is generally defined as the distance between the upper and the lower layer of the fouling interface. The value is usually expressed in pixel and then converted in μm taking into account the refractive index of the penetrated media [26]. In literature, it is assumed a single refractive index equal to 1.333. The other key descriptors extracted from the image analysis are related to the fouling homogeneity. Relative and absolute roughnesses are considered as roughness measurements where the first one is dimensionless and the second one is in μm [26, 36]. The macroporosity was also introduced as parameters to quantify the presence of big voids inside the fouling structure [36]. Another important value calculated from the analysis is the membrane coverage that expresses the amount of membrane covered by fouling [37].

3.2 Monitoring the fouling growth under continuous operation

Due to its ability to monitor the system without stopping or affecting in any ways, OCT process has been used so far in several membrane filtrations systems and configurations. The central advantage respect to the other monitoring techniques is represented by the possibility of "having a window on the system" (**Figure 6**), and thus being able to correlate the decrease in performances to the in-situ investigation. Among the different types of membrane configurations, the flat sheet membrane bioreactor is one of the easiest systems on which the OCT is employed. Flat sheet membranes are easy to install and compared to the hollow fiber, they have the advantage of being flat, therefore, keeping a fixed distance from the membrane to the OCT probe. This enables covering bigger areas on a single scan moving easily among different positions in the membrane module. The research in submerged system representative of reactors is performed on tank of Plexiglas, where the membrane is submerged. OCT was employed to evaluate the change of morphology during 42 days in a gravity-driven membrane bioreactor treating synthetic wastewater [36]. As shown in **Figure 7**, a significative change in morphology was observed over time, highlighting the evolution and the change of the fouling layer during the operation.

The possibility of capturing a fouling morphology change has been exploited by several authors. Some tests were focused on evaluating the change of morphologies linked to the change of the feed [38, 39]. Experiments are often performed in membrane flowcells and compared to a control. Shao et al. [40] compared the backwashing efficiency in controlling the biofouling by using two different backwash feeds. Derlon et al. [26] used the OCT to evaluate the effect of metazoan as biological control of biofilm developed on the membrane. The use of metazoan led to a change in fouling morphology that was also linked to an enhancement in the flux. Farid et al. [41] used the OCT to evaluate the bacterial inactivation of a graphene oxide

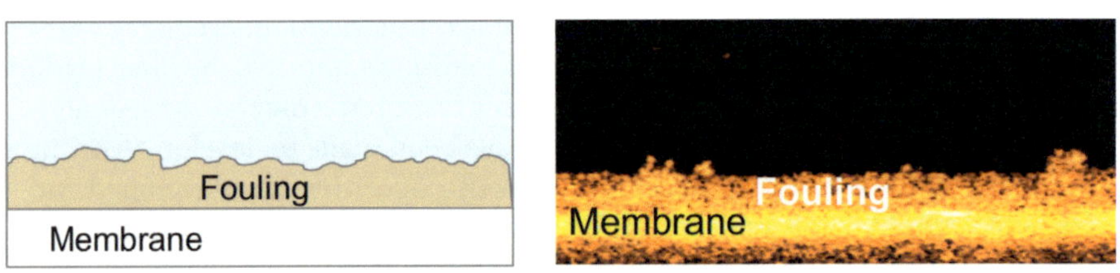

Figure 6.
Schematic representation of OCT in-situ monitoring in a membrane filtration process. 2D OCT scan acquired under continuous operation of the biomass deposited on the membrane surface.

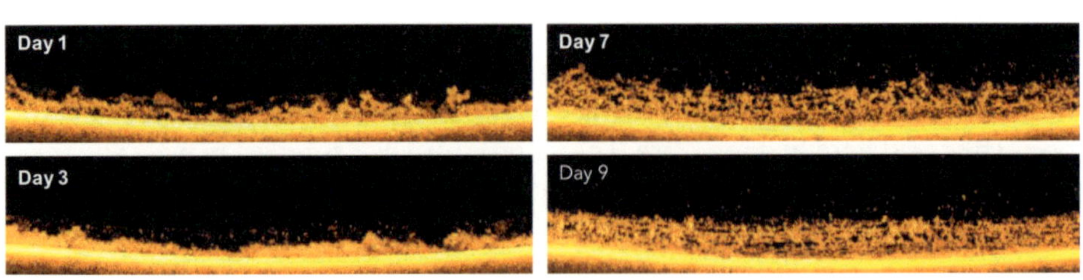

Figure 7.
Time-resolved analysis of fouling deposited on a flat sheet membrane treating wastewater. Adapted from Fortunato et al. [34].

membrane. In this case, the use of the OCT highlighted the difference in deposition between a commercial membrane and a graphene oxide membrane with antifouling proprieties. The OCT was employed to study the compression and decompression of the structure were observed and to relate to the biofilm mechanical proprieties [42, 43]. The authors observed an increase in pressure drop and hydraulic resistance over time. Desmond et al. [44] employed the OCT to study the compression of membrane biofilms in gravity-driven ultrafiltration. OCT scans were also used by Wibisono et al. [45] to evaluate the efficacy of two-phase cleaning flow in a spiral-wound element. Recently, the OCT has also been employed to validate a fluorescence-based method for the detection of biofouling at the early stage [46].

Another interesting benefit of using OCT in filtration processes is represented by the opportunity of recording the cross section at high frequency. Blauert et al. [47] used this approach to evaluate the time-resolved deformation of a biofilm in a flow cell enabling the estimation of the mechanical proprieties. Fortunato et al. [37], by using time-resolved analysis, were able to generate a video of the biomass developed during early stage filtration on a submerged membrane. The image analysis performed on the scans showed a correlation between the biomass membrane coverage and the biomass thickness with the decrease of fouling at an early stage. Moreover, through the videos, it was possible to correlate the biomass development with the flux decrease and capture particular morphologies constituted by a double-layer structure, where the upper one was moving and the lower one was still.

3.3 OCT monitoring in spacer-filled channel

The same approach was extended to spacer-filled channel spiral wound module, which is the membrane module employed for desalination. In spiral-wound membrane, biofouling has been identified as the bottleneck [48], since it leads to decrease in performances in the full-scale operating plants due to the increase in feed channel pressure drop, permeate flux reduction, and/or salt passage increase [49]. The spacer-filled channel geometry is a characteristic of spiral wound elements, where the spacer is used to increase the turbulence and separate the membrane sheets. This configuration is considered more complex for imaging purpose due to the presence of a plastic feed spacer that complicates the image processing and morphology analysis. The research in this field is carried out by using a flow cell called membrane fouling simulator (MFS) representative of the hydrodynamics, and it has all the elements present in the module. In monitoring the fouling in spacer-filled channel, the ability of the OCT in acquiring scans in the mm range resulted even essential due to the presence of the feed spacer. In fact, by using the OCT, it was possible to monitor the biomass deposited on the feed spacer pattern representative of the hydrodynamics of the module. West et al. [50] performed image analysis on 3D OCT dataset on two different feed spacer meshes. Fortunato et al. [33] visualized and quantified the 3D structure of the biomass deposited in the system, enabling to assess the spatial distribution of the biomass in the channel, whereas the highest deposition was observed on the feed spacer. Afterward, the biofilm thickness map was proposed as a tool to quickly evaluate the biomass deposited in spacer-filled channel (**Figure 8**) [34].

3.4 Monitoring inorganic fouling

In the case of membrane filtration systems, the OCT was at the beginning employed to study the biofilm and the biofouling formation, afterward the in-situ monitoring was also extended to other types of fouling. Online monitoring was

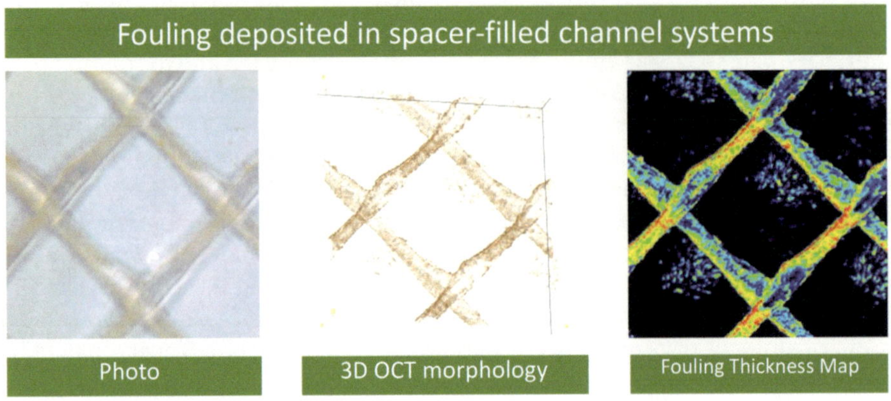

Figure 8.
Fouling characterization in spacer-filled channel systems through 3D OCT image analysis. The spacer-filled channel geometry is typical of the membrane employed for sweater desalination. Adapted from Fortunato et al. [34].

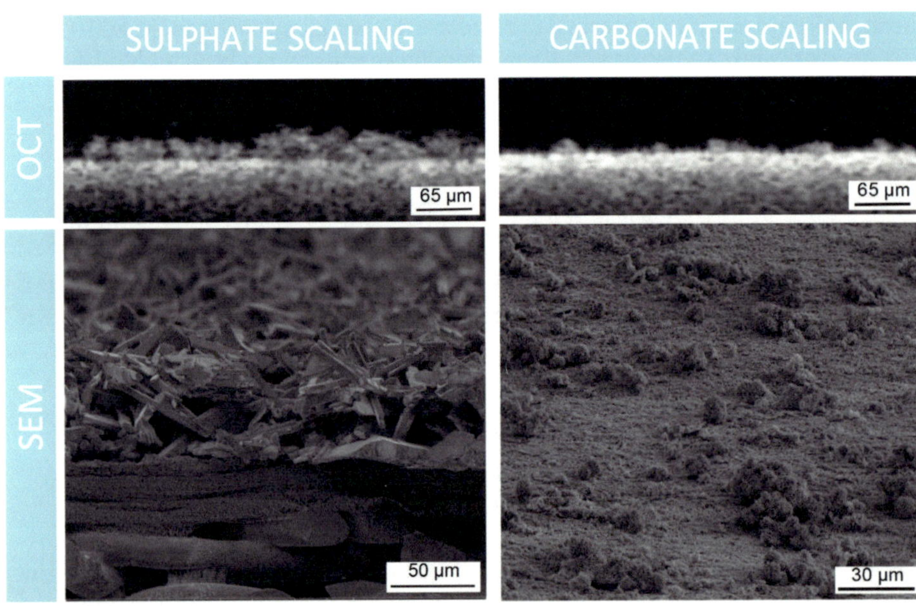

Figure 9.
Inorganic fouling structure deposited on membrane surface treating saline feed. OCT scans and SEM images. Adapted from Fortunato et al. [52].

performed on fouling layer formed by different particles as silica and bentonite [32, 51]. Recently, Fortunato et al. [52] employed the OCT to monitor the formation of scaling in a membrane distillation process. Membrane distillation is a hybrid process that couples thermal and membrane processes used to treat high saline feed as the brine. In these studies, the OCT analysis was coupled with a membrane autopsy to identify the nature of the deposition. Thought the in-situ analysis was possible to evaluate the formation of carbonate and sulfate crystals over time (**Figure 9**). The use of OCT was then proposed as a tool to monitor the scaling in thermal process that employs membrane [53]. Recently, Bauer et al. [54] used the OCT to quantify the area covered by the scaling and the flux decline in a membrane distillation process. In 2018, the OCT was employed to analyze the external and internal fouling due to oil droplets [55].

3.5 Using the OCT to improve fluid dynamic simulation

One of the main objectives of the fouling structural analysis is to evaluate the impact of the biomass deposited on the membrane on the performance. Modeling is often performed on filtration process to evaluate the effect of the shear force on the biomass formed and provide a better understanding of the process. Furthermore, there has always been a demand to implement the biomass structure in multidimensional models to understand the structure-fluid interaction and predict the behavior. Indeed, a real biomass structure better matches with the process performance respect to a theoretical structure with a given average thickness. Initially, the acquisition of the structure that affected the membrane permeability was performed by means of the CLSM [56]; however, as stated in Section 2, the CLSM has several disadvantages with respect to the OCT, including the incapability of acquiring data nondestructively.

Moreover, OCT enables to acquire information at wider scale allowing to study an area more representative of the process. The OCT scans were used by Martin et al. [57] to perform simulation of the permeate flux in a gravity driven system. Gao et al. [58] used the Doppler effect to visualize the velocity field in a spacer-filled channel. Fortunato et al. [36] imported the real biomass morphologies developed on a membrane bioreactor-treated secondary wastewater effluent. The biomass morphology was imported after 3 and 30 days of filtration and implemented in a computational fluid dynamic simulation (CFD), allowing to identify the local region of local and high flux within the biomass structure. The approach proposed allowed to match the model with the experimental values of the permeate fluxes. Jafari et al. [59] developed a numerical model able to correlate the structural deformation with biofilm hydraulics by using the in-situ observation performed with OCT. Recently, Picioreanu et al. [60] employed the OCT scans to develop a method for the determination of the elastic proprieties of a biofilm. In summary, coupling the OCT with the CFD represents a powerful toolbox to understand and predict the behavior of the biomass in membrane filtration processes.

4. Conclusions

Membrane fouling is considered the main limitation of membrane filtration systems in terms of cost and operation. The techniques commonly used for analyzing the fouling are based on membrane autopsies, where membrane coupons were collected and analyzed after destroying a membrane module. With that approach, therefore, it is possible to provide information only at the specific time chosen to conduct the autopsy. Those techniques are consequently subject to the circumstance of ending the process. Moreover, some of the techniques necessitate sampling preparation such as drying or labeling, which have the risk of changing the fouling morphology. Therefore, considering that the fouling is a dynamic process and will evolve, it is necessary to monitor the fouling development over time under continuous operation without interfering with the process.

Thanks to the possibility of monitoring samples without the use of staining in-situ nondestructively, the use of OCT gained attention in studying the fouling in membrane filtration systems. At the beginning, the OCT was employed to study the biofilm formed in the process and later applied to all the different types of fouling. Nowadays, the OCT is considered an essential tool to gain a better understanding of fouling behavior and is employed in different membrane configurations and systems covering the whole spectrum of membrane filtration processes. The in-situ nondestructive online acquisition cross-sectional scans of the fouling deposited enables to link the impact of the fouling on the membrane performance (i.e., flux decrease and feed channel pressure drop). The approach also resulted to be beneficial in evaluating the efficacy of antifouling strategies. In summary, the use of OCT in membrane filtration systems turned out to be a key tool in understanding and predicting the fouling development and its effect on the overall membrane performance.

Acknowledgements

This study was supported by funding from King Abdullah University of Science and Technology (KAUST).

Conflict of interest

I confirm there are no conflicts of interest.

Author details

Luca Fortunato
Division of Biological and Environmental Science and Engineering (BESE), Water Desalination and Reuse Center (WDRC), King Abdullah University of Science and Technology (KAUST), Thuwal, Saudi Arabia

*Address all correspondence to: luca.fortunato@kaust.edu.sa

References

[1] Rost S, Gerten D, Bondeau A, Lucht W, Rohwer J, Schaphoff S. Agricultural green and blue water consumption and its influence on the global water system. Water Resources Research. 2008;**44**. DOI: 10.1029/2007WR006331

[2] U.& WWAP. The United Nations World Water Development Report 2;2006

[3] Baker RW. Membrane Technology and Applications2012. DOI: 10.1002/9781118359686

[4] Drews A. Membrane fouling in membrane bioreactors—Characterisation, contradictions, cause and cures. Journal of Membrane Science. 2010;**363**:1-28. DOI: 10.1016/j.memsci.2010.06.046

[5] Le-Clech P, Chen V, Fane TAG. Fouling in membrane bioreactors used in wastewater treatment. Journal of Membrane Science. 2006;**284**:17-53. DOI: 10.1016/j.memsci.2006.08.019

[6] Nguyen T, Roddick FA, Fan L. Biofouling of water treatment membranes: A review of the underlying causes, monitoring techniques and control measures. Membranes (Basel). 2012;**2**:804-840. DOI: 10.3390/membranes2040804

[7] Flemming HC. Biofouling in water systems—Cases, causes and countermeasures. Applied Microbiology and Biotechnology. 2002;**59**:629-640. DOI: 10.1007/s00253-002-1066-9

[8] Flemming HC, Schaule G, Griebe T, Schmitt J, Tamachkiarowa A. Biofouling—The Achilles heel of membrane processes. Desalination. 1997;**113**:215-225. DOI: 10.1016/S0011-9164(97)00132-X

[9] Weinrich L, Haas CN, LeChevallier MW. Recent advances in measuring and modeling reverse osmosis membrane fouling in seawater desalination: A review. Journal of Water Reuse and Desalination. 2013;**3**:85. DOI: 10.2166/wrd.2013.056

[10] Halan B, Buehler K, Schmid A. Biofilms as living catalysts in continuous chemical syntheses. Trends in Biotechnology. 2012;**30**:453-465. DOI: 10.1016/j.tibtech.2012.05.003

[11] Valladares Linares R, Fortunato L, Farhat NM, Bucs SS, Staal M, Fridjonsson EO, et al. Mini-review: Novel non-destructive in situ biofilm characterization techniques in membrane systems. Desalination and Water Treatment. 2016;**57**:22894-22901. DOI: 10.1080/19443994.2016.1180483

[12] Fortunato L, Jeong S, Wang Y, Behzad AR, Leiknes T. Integrated approach to characterize fouling on a flat sheet membrane gravity driven submerged membrane bioreactor. Bioresource Technology. 2016;**222**:335-343. DOI: 10.1016/j.biortech.2016.09.127

[13] Lawrence JR, Swerhone GDW, Leppard GG, Araki T, Zhang X, West MM, et al. Scanning transmission X-ray, laser scanning, and transmission electron microscopy mapping of the exopolymeric matrix of microbial biofilms. Applied and Environmental Microbiology. 2003;**69**:5543-5554. DOI: 10.1128/AEM.69.9.5543-5554.2003

[14] Herzberg M, Elimelech M. Biofouling of reverse osmosis membranes: Role of biofilm-enhanced osmotic pressure. Journal of Membrane Science. 2007;**295**:11-20. DOI: 10.1016/j.memsci.2007.02.024

[15] Flemming HC, Wingender J. The biofilm matrix. Nature Reviews. Microbiology. 2010;**8**:623-633. DOI: 10.1080/0892701031000072190

[16] Lewandowski Z, Webb D, Hamilton M, Harkin G. Quantifying biofilm structure. Water Science and Technology. 1999;**39**:71-76. DOI: 10.1016/S0273-1223(99)00152-3

[17] Jin X, Shan J, Wang C, Wei J, Tang CY. Rejection of pharmaceuticals by forward osmosis membranes. Journal of Hazardous Materials. 2012;**227-228**:55-61. DOI: 10.1016/j.jhazmat.2012.04.077

[18] Wagner M, Taherzadeh D, Haisch C, Horn H. Investigation of the mesoscale structure and volumetric features of biofilms using optical coherence tomography. Biotechnology and Bioengineering. 2010;**107**:844-853. DOI: 10.1002/bit.22864

[19] Graf von der Schulenburg DA, Vrouwenvelder JS, Creber SA, van Loosdrecht MCM, Johns ML. Nuclear magnetic resonance microscopy studies of membrane biofouling. Journal of Membrane Science. 2008;**323**:37-44. DOI: 10.1016/j.memsci.2008.06.012

[20] Fridjonsson EO, Vogt SJ, Vrouwenvelder JS, Johns ML. Early non-destructive biofouling detection in spiral wound RO membranes using a mobile earth's field NMR. Journal of Membrane Science. 2015;**489**:227-236. DOI: 10.1016/j.memsci.2015.03.088

[21] Frederiksen MS, Glud RN. Oxygen dynamics in the rhizosphere of Zostera marina: A two-dimensional planar optode study. Limnology and Oceanography. 2006;**51**:1072-1083

[22] Wojtkowski M, Srinivasan V, Fujimoto JG, Ko T, Schuman JS, Kowalczyk A, et al. Three-dimensional retinal imaging with high-speed ultrahigh-resolution optical coherence tomography. Ophthalmology. 2005;**112**:1734-1746. DOI: 10.1016/j.ophtha.2005.05.023

[23] Xi C, Marks D, Schlachter S, Luo W, Boppart S a. High-resolution three-dimensional imaging of biofilm development using optical coherence tomography. Journal of Biomedical Optics. 2006;**11**:034001. DOI: 10.1117/1.2209962

[24] Haisch C, Niessner R. Visualisation of transient processes in biofilms by optical coherence tomography. Water Research. 2007;**41**:2467-2472. DOI: 10.1016/j.watres.2007.03.017

[25] Wagner M, Horn H. Optical coherence tomography in biofilm research: A comprehensive review. Biotechnology and Bioengineering. 2017;**114**:1386-1402. DOI: 10.1002/bit.26283

[26] Derlon N, Peter-Varbanets M, Scheidegger A, Pronk W, Morgenroth E. Predation influences the structure of biofilm developed on ultrafiltration membranes. Water Research. 2012;**46**:3323-3333. DOI: 10.1016/j.watres.2012.03.031

[27] Dreszer C, Flemming HC, Zwijnenburg A, Kruithof JC, Vrouwenvelder JS. Impact of biofilm accumulation on transmembrane and feed channel pressure drop: Effects of crossflow velocity, feed spacer and biodegradable nutrient. Water Research. 2014;**50**:200-211. DOI: 10.1016/j.watres.2013.11.024

[28] Fortunato L, Li M, Cheng T, Rehman ZU, Heidrich W, Leiknes T. Cake layer characterization in activated sludge membrane bioreactors: Real-time analysis. Journal of Membrane Science. 2019;**578**:163-171. DOI: 10.1016/J.MEMSCI.2019.02.026

[29] BeyenalH,DonovanC,LewandowskiZ, Harkin G. Three-dimensional biofilm structure quantification. Journal of Microbiological Methods. 2004;**59**: 395-413. DOI: 10.1016/J.MIMET.2004.08.003

[30] Li M, Idoughi R, Choudhury B, Heidrich W. Statistical model for OCT

image denoising. Biomedical Optics Express. 2017;**8**:3903. DOI: 10.1364/BOE.8.003903

[31] Fortunato L, Pathak N, Ur Rehman Z, Shon H, Leiknes T. Real-time monitoring of membrane fouling development during early stages of activated sludge membrane bioreactor operation. Process Safety and Environment Protection. 2018;**120**:313-320. DOI: 10.1016/J.PSEP.2018.09.022

[32] Gao Y, Haavisto S, Li W, Tang CY, Salmela J, Fane AG. Novel approach to characterizing the growth of a fouling layer during membrane filtration via optical coherence tomography. Environmental Science and Technology. 2014;**48**:14273-14281. DOI: 10.1021/es503326y

[33] Fortunato L, Bucs S, Linares RV, Cali C, Vrouwenvelder JS, Leiknes T. Spatially-resolved in-situ quantification of biofouling using optical coherence tomography (OCT) and 3D image analysis in a spacer filled channel. Journal of Membrane Science. 2017;**524**:673-681. DOI: 10.1016/j.memsci.2016.11.052

[34] Fortunato L, Leiknes TO. In-situ biofouling assessment in spacer filled channels using optical coherence tomography (OCT): 3D biofilm thickness mapping. Bioresource Technology. 2017;**229**:231-235. DOI: 10.1016/j.biortech.2017.01.021

[35] Li W, Liu X, Wang Y-N, Chong TH, Tang CY, Fane AG. Analyzing the evolution of membrane fouling via a novel method based on 3D optical coherence tomography imaging. Environmental Science and Technology. 2016;**50**:6930-6939. DOI: 10.1021/acs.est.6b00418

[36] Fortunato L, Qamar A, Wang Y, Jeong S, Leiknes T. In-situ assessment of biofilm formation in submerged membrane system using optical coherence tomography and computational fluid dynamics. Journal of Membrane Science. 2017;**521**:84-94. DOI: 10.1016/j.memsci.2016.09.004

[37] Fortunato L, Jeong S, Leiknes T. Time-resolved monitoring of biofouling development on a flat sheet membrane using optical coherence tomography. Scientific Reports. 2017;**7**:15. DOI: 10.1038/s41598-017-00051-9

[38] Pathak N, Fortunato L, Li S, Chekli L, Phuntsho S, Ghaffour N, et al. Evaluating the effect of different draw solutes in a baffled osmotic membrane bioreactor-microfiltration using optical coherence tomography with real wastewater. Bioresource Technology. 2018;**263**:306-316. DOI: 10.1016/j.biortech.2018.04.123

[39] Wang Y, Fortunato L, Jeong S, Leiknes TO. Gravity-driven membrane system for secondary wastewater effluent treatment: Filtration performance and fouling characterization. Separation and Purification Technology. 2017;**184**:26-33. DOI: 10.1016/j.seppur.2017.04.027

[40] Shao S, Wang Y, Shi D, Zhang X, Tang CY, Liu Z, et al. Biofouling in ultrafiltration process for drinking water treatment and its control by chlorinated-water and pure water backwashing. Science of the Total Environment. 2018;**644**:306-314. DOI: 10.1016/J.SCITOTENV.2018.06.220

[41] Farid MU, Guo J, An AK. Bacterial inactivation and in situ monitoring of biofilm development on graphene oxide membrane using optical coherence tomography. Journal of Membrane Science. 2018;**564**:22-34. DOI: 10.1016/J.MEMSCI.2018.06.061

[42] Valladares Linares R, Wexler AD, Bucs SS, Dreszer C, Zwijnenburg A, Flemming HC, et al. Compaction and relaxation of biofilms. Desalination and Water Treatment. 2016;**57**:12902-12914. DOI: 10.1080/19443994.2015.1057036

[43] Dreszer C, Wexler AD, Drusová S, Overdijk T, Zwijnenburg A, Flemming HC, et al. In-situ biofilm characterization in membrane systems using optical coherence tomography: Formation, structure, detachment and impact of flux change. Water Research. 2014;**67**:243-254. DOI: 10.1016/j.watres.2014.09.006

[44] Desmond P, Morgenroth E, Derlon N. Physical structure determines compression of membrane biofilms during gravity driven membrane (GDM) ultrafiltration. Water Research. 2018;**143**:539-549. DOI: 10.1016/J.WATRES.2018.07.008

[45] Wibisono Y, El Obied KE, Cornelissen ER, Kemperman AJB, Nijmeijer K. Biofouling removal in spiral-wound nanofiltration elements using two-phase flow cleaning. Journal of Membrane Science. 2015;**475**:131-146. DOI: 10.1016/j.memsci.2014.10.016

[46] Khan BK, Fortunato L, Leiknes T. Early biofouling detection using fluorescence-based extracellular enzyme activity. Enzyme and Microbial Technology. 2019;**120**:43-51. DOI: 10.1016/J.ENZMICTEC.2018.10.001

[47] Blauert F, Horn H, Wagner M. Time-resolved biofilm deformation measurements using optical coherence tomography. Biotechnology and Bioengineering. 2015;**112**:1893-1905. DOI: 10.1002/bit.25590

[48] Vrouwenvelder JS, Bakker SM, Wessels LP, van Paassen JAM. The membrane fouling simulator as a new tool for biofouling control of spiral-wound membranes. Desalination. 2007;**204**:170-174. DOI: 10.1016/j.desal.2006.04.028

[49] Characklis W, Marshall K. Biofilms. New York: John Wiley & Sons; 1990

[50] West S, Wagner M, Engelke C, Horn H. Optical coherence tomography for the in situ three-dimensional visualization and quantification of feed spacer channel fouling in reverse osmosis membrane modules. Journal of Membrane Science. 2015;**498**:345-352. DOI: 10.1016/j.memsci.2015.09.047

[51] Li C, Wagner M, Lackner S, Horn H. Assessing the influence of biofilm surface roughness on mass transfer by combining optical coherence tomography and two-dimensional modeling. 2016;**113**:989-1000. DOI: 10.1002/bit.25868

[52] Fortunato L, Jang Y, Lee JG, Jeong S, Lee S, Leiknes TO, et al. Fouling development in direct contact membrane distillation: Non-invasive monitoring and destructive analysis. Water Research. 2018;**132**:34-41. DOI: 10.1016/j.watres.2017.12.059

[53] Lee JG, Jang Y, Fortunato L, Jeong S, Lee S, Leiknes T, et al. An advanced online monitoring approach to study the scaling behavior in direct contact membrane distillation. Journal of Membrane Science. 2018;**546**:50-60. DOI: 10.1016/J.MEMSCI.2017.10.009

[54] Bauer A, Wagner M, Saravia F, Bartl S, Hilgenfeldt V, Horn H. In-situ monitoring and quantification of fouling development in membrane distillation by means of optical coherence tomography. Journal of Membrane Science. 2019;**577**:145-152. DOI: 10.1016/J.MEMSCI.2019.02.006

[55] Trinh TA, Li W, Han Q, Liu X, Fane AG, Chew JW. Analyzing external and internal membrane fouling by oil emulsions via 3D optical coherence tomography. Journal of Membrane Science. 2018;**548**:632-640. DOI: 10.1016/J.MEMSCI.2017.10.043

[56] Horn H, Lackner S. Modeling of biofilm systems: A review. In: Muffler K, Ulber R, editors. Productive Biofilms. Cham: Springer International Publishing; 2014. pp. 53-76. DOI: 10.1007/10_2014_275

[57] Martin KJ, Bolster D, Derlon N, Morgenroth E, Nerenberg R. Effect of fouling layer spatial distribution on permeate flux: A theoretical and experimental study. Journal of Membrane Science. 2014;**471**:130-137. DOI: 10.1016/j.memsci.2014.07.045

[58] Gao Y, Haavisto S, Tang CY, Salmela J, Li W. Characterization of fluid dynamics in spacer-filled channels for membrane filtration using Doppler optical coherence tomography. Journal of Membrane Science. 2013;**448**:198-208. DOI: 10.1016/j.memsci.2013.08.011

[59] Jafari M, Desmond P, van Loosdrecht MCM, Derlon N, Morgenroth E, Picioreanu C. Effect of biofilm structural deformation on hydraulic resistance during ultrafiltration: A numerical and experimental study. Water Research. 2018;**145**:375-387. DOI: 10.1016/J.WATRES.2018.08.036

[60] Picioreanu C, Blauert F, Horn H, Wagner M. Determination of mechanical properties of biofilms by modelling the deformation measured using optical coherence tomography. Water Research. 2018;**145**:588-598. DOI: 10.1016/J.WATRES.2018.08.070

Chapter 7

Nondestructive Characterization of Drying Processes of Colloidal Droplets and Latex Coats Using Optical Coherence Tomography

Yongyang Huang, Hao Huang, Zhiyu Jiang, Lanfang Li, Willie Lau, Mohamed El-Aasser, Hsin-Chiao Daniel Ou-Yang and Chao Zhou

Abstract

In this chapter, we review the applications of optical coherence tomography (OCT) on the nondestructive characterization of the drying processes of colloidal droplets and latex coatings. Employing time-lapse, high-speed imaging, OCT can be used to monitor the dynamic process of drying colloidal droplets. With the aid of high-scattering, micron-sized tracer particles, fluid flows have been captured; phase boundaries are also visible in liquid crystal droplets; and the speckle contrast analysis differentiates the dynamics of particles, showing the packing process and the coffee ring phenomenon. In a waterborne latex coat, time-lapse OCT imaging reveals spatial changes of microstructures, i.e., detachment of latex, cracks, and shear bands; with speckle contrast analysis, 1D and 2D particles' packing process that is initiated from latex/air interface can also be monitored over time. OCT can serve as an experimental platform for fundamental studies of drying colloidal systems. In the future, OCT can also be employed as an in-line quality control tool of polymer coatings and paints for industrial applications.

Keywords: optical coherence tomography, drying, colloidal droplet, waterborne latex, liquid crystal droplet, time-lapse imaging, speckle

1. Introduction

In this chapter, we describe the applications of optical coherence tomography (OCT) on the characterization of drying dynamics of colloidal systems. Specifically, we will use OCT in combination with other modalities (i.e., gravimetry and video recording) to characterize two drying models: (1) drying colloidal droplets and (2) drying latex coat. The entire drying processes of these systems can take from a few minutes for microliter droplets to a few hours or days for thick-film latex in Petri dishes. Given the high-speed imaging capability with a temporal resolution on the order of milliseconds, OCT can detect the flow of fluid or particle motions. Given the nondestructive nature of OCT, time-lapse OCT can be used to monitor the drying processes of droplets and latex coats.

2. Drying process of colloidal droplets

Drying of the colloidal droplets has been a subject of great interest since the 1980s [1]. Studies of a drying droplet, including changes of drying rate and contact angle, the progression of its shape, and final deposition pattern, can help us understand many interesting phenomena, such as coffee ring effects [2], electro-wetting effects [3], and Marangoni effects that are driven by surface tension gradients [4]. Mechanisms behind these phenomena may involve a complex interplay of convection and evaporation, surface tension and capillary force, particles' interactions, rheology, substrate hydrophobicity, pinning of the contact line, and Marangoni forces [3, 5]. During drying, unlike pure water droplets, the constituents in colloidal droplets, including various concentrations of particles [6], polymers [7–9], surfactants [10], solvents [11], and salts [12] will alter the interactions among these mass transfer mechanisms, yielding different drying behaviors. Additionally, environmental conditions such as temperature and humidity and the substrate properties also affect the drying process and final deposition [13, 14]. Insights gained from these studies of evaporating droplets may ultimately translate to practical applications in polymer science, biomedicine, and nanotechnology [14], such as inkjet printing [15], DNA chip [16], biosensor [17, 18] and disease diagnosis based on deposition pattern of biological fluid [19–22], food quality analysis [23], particle separation [24], and production of nanoparticles for drug delivery [25, 26]. All above-mentioned applications are related to mass transportation during drying of a liquid suspension and the resultant film property.

Using OCT to monitor the drying process of colloidal droplets has been conducted in a few studies. **Table 1** summarizes the details of experimental designs for these studies. In conjunction, **Figure 1** illustrates the schematics of the drying process of these colloidal droplet models.

Trantum et al. demonstrated the first study to utilize OCT to visualize the cross-section of drying water droplets with suspended particles (see **Figure 1A**) [27]. Low concentration of polystyrene, melamine formaldehyde, or silica particles with a diameter of 1 μm and a volume fraction of 0.005% was suspended in distilled water. The water droplet was loaded on a hydrophilic glass slide. The drying experiment was conducted in the ambient condition with relative humidity (RH) controlled to 30 or 40%. A commercial spectral-domain OCT system with a central wavelength of 860 nm and a spectral bandwidth of 51 nm was used to image the droplet. Axial and lateral resolutions of the OCT system were 6.4 and 8 μm, respectively. The scan rate of the system was set to 10 kHz. Their results showed that both the sedimentation rate and evaporation rate would affect the drying dynamics of particle-containing water droplets. In fast evaporating mode, the descending rate of drop surface was faster than the rate of particle sedimentation, resulting in the particles being trapped at the air-water interface (top surface). Close to the water-substrate interface, the "coffee ring" flows (CF) played a major role to transport the particles to the pinned contact line, yielding a "coffee ring" final deposition pattern. Given the same particle size, increased particle densities, i.e., silica (2.00 g/cm^3), as compared to polystyrene (1.04 g/cm^3) lead to early sediment of particles and early transportation of particles to the edge by "coffee ring" flows. Their results suggested a way to control the final deposition of coffee ring structure based on particles' density.

Based on the previous observations of drying colloidal droplets, the same group proposed design of biosensor utilizing the final deposition pattern to track the existence and concentration of target biomarkers (**Figure 1B**) [17]. In the experiment, the tracer particles were coated with antibodies. These particles would aggregate in the presence of target biomarker (M13 bacteriophage in the study). Polydimethylsiloxane (PDMS) with low thermal conductivity (0.15 W/mK) was

References	Trantum et al. [27]	Trantum et al. [17]	Manukyan et al. [28]	Davidson et al. [3]	Edwards et al. [29]	Huang et al. [30]
Years	2013	2014	2013	2017	2018	2018
Suspended materials	PS, MF, Si*	Monoclonal antibodies Glycerol with M13**	Model paint	LCLC, SSF***	Binary drops (ethanol and n-butanol)	PS Latex
Diameter (nm)	1000	1000	<1000	N.A.	N.A.	L Latex 125 S Latex 53
Initial concentration	vol%: 0.005% (10^5/μL)	vol%: 0.05% (10^6/μl) Glycerol: 8%	vol%: 10, 30, 50%	wt%: 5–20%	wt%: 2–20%	wt%: L latex: 40.11%, S latex: 33.43%
Substrate	Glass slide	PDMS	Glass coated by Tegotop® 210	Glass slide and coverslip	Glass coated by Flutech LE15	Glass slide
Environment	Ambient	Ambient	Ambient	Chambered	Chambered	Ambient
Droplet model	Sessile	Sessile	Sessile	Sessile	Sessile or pendant	Sessile
Transparency****	Transparent	Transparent	Opaque	Transparent	Transparent	Opaque
Tracers	None		Copolymer microspheres diameter: 4.3 or 7.9 μm	Polystyrene diameter: 1 μm	0.01 wt% polystyrene diameter: 2 μm	None
OCT type	Spectral domain		Spectral domain	Spectral domain	N.A.	Spectral domain
λ_{OCT} (nm)	860		930	800	1300	1320
$\Delta\lambda_{OCT}$ (nm)	51		100	220	N.A.	110
Axial and lateral resolutions (in air)	6.4 μm 8 μm		7 μm 9 μm	1.9 μm 3.5 μm	5.5 μm 13 μm	6.8 μm 14 μm
Scan speed (kHz)	10		N.A.	20	N.A.	20.7
Imaging depth and scan range	N.A.		1.6 mm 4 mm	N.A.	N.A. 2.5 mm	2.2 mm 5.0 mm

References	Trantum et al. [27]	Trantum et al. [17]	Manukyan et al. [28]	Davidson et al. [3]	Edwards et al. [29]	Huang et al. [30]
Frame rate or time interval	5 or 0.5 fps, 200 frames		2 fps	33 or 50 fps interval: ~3–4 s	50 fps	Interval: ~10 s

**PS: polystyrene, MF: melamine formaldehyde, Si: silica.*
***Particle surface was functionalized with anti-M13.*
****LCLC: lyotropic chromonic liquid crystal. SSF: sunset yellow FCF.*
*****Optical transparency in near-IR wavelength range.*

Table 1.
OCT studies on drying droplets.

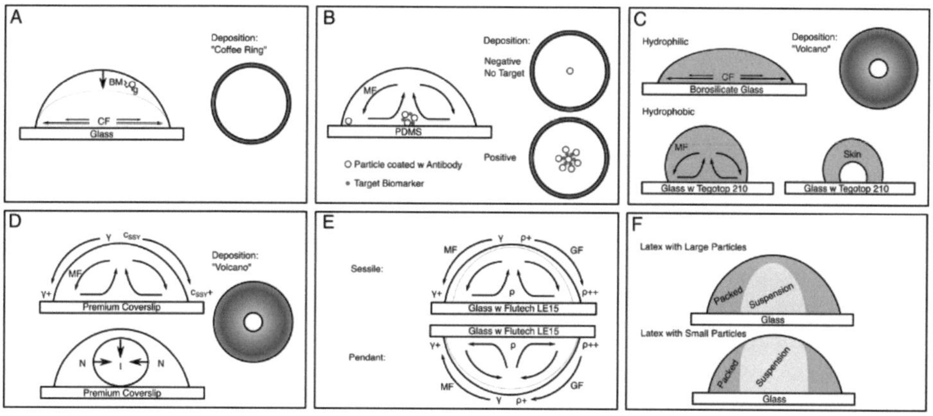

Figure 1.
Illustrations of drying process of six colloidal droplet models. (A) Water droplets with tracer particles, i.e., polystyrene, melamine formaldehyde, silica. (B) Water droplet with antibody-coated particles to detect biomarkers. (C) Model paint droplet. (D) Liquid crystal droplet. (E) Binary liquid droplet. (F) Waterborne latex droplet.

used as the substrate. Thus, a temperature gradient was established along with the interface, inducing a surface tension gradient and promoting Marangoni flows. During the drying process, the particle aggregates in the presence of biomarkers showed a thicker convective flow pattern in the cross-sectional OCT images, while the dispersed particles showed a uniformly distributed convective flow pattern in the absence of biomarkers [17]. Under the influence of Marangoni flows, increase of viscosity by glycerol, and effect of gravity, the particle aggregates were concentrated and deposited at the droplet center, forming a concentric final deposition pattern. Note that, with the glass substrate, the particle aggregates will not concentrate at the droplet center.

Manukyan et al. reported using OCT to characterize the internal flows in a drying model paint droplet (**Figure 1C**) [28]. Commercial model paint droplets with various initial volume fractions were characterized. Copolymer microspheres were mixed in the model paint droplets as tracer particles. Model paint droplets with an initial volume of 5–8 μL were loaded on hydrophilic or hydrophobic substrates, and their drying behaviors were monitored separately. A spectral-domain OCT system was utilized to perform time-lapse imaging of model paint droplets. The central wavelength and the spectral bandwidth of the system were 930 and 100 nm. Axial and lateral resolutions of the OCT system were 7 and 9 μm, respectively. The OCT system can scan a maximum transverse range of 4 mm. In their results, the model paint droplets on a hydrophilic substrate exhibited an outward radial flow pattern and a donut-shaped final deposition pattern in cross-sectional OCT images, indicating that the drying processes of these droplets were driven mainly by the coffee ring effect. However, on the hydrophobic surface, a reversed Marangoni flow was established in the drying of model paint droplets, shown as the convective flow moving down along the surface and then moving inward along the droplet-substrate interface in the OCT image. As the model paint droplet continued to dry, the formation of a skin layer was observed inside the droplet. At this stage, no convective flows were observed. A cavity was observed beneath the outer skin layer in OCT cross-sectional images.

Davidson et al. investigated the drying process of water droplets containing lyotropic chromonic liquid crystals (LCLCs), using polarized optical microscopy (POM) and OCT (**Figure 1D**) [3]. LCLCs are composed of organic, charged, and plank-like molecules [31, 32]. During the drying process of LCLC containing droplets, changes of local temperature and concentration of LCLCs induce multiple phases of the LCLCs, including isotropic, nematic, columnar, and crystalline phases [3]. In the experiment, a droplet containing sunset yellow FCF (SSY), a dye that belonged to LCLC family, was loaded on a premium coverslip of the substrate. Polystyrene particles were added in the droplets as tracers to track fluid flows. A semi-enclosed PDMS chamber was used to slow down the drying rate, yielding a total drying time of ~10–15 min for the LCLC containing droplets with an initial volume of ~0.2–0.5 μL. To visualize the small tracer particles in the LCLC containing droplet, an ultrahigh-resolution (UHR) spectral-domain OCT system was employed, with a central wavelength of 800 nm and a spectral bandwidth of 220 nm. Axial and lateral resolutions of the UHR-OCT system were 1.9 μm and 3.5 μm, respectively. The camera's scan rate was set to 20 kHz. **Figure 2** shows the development of drying SSY solution droplet by UHR-OCT. High-speed time-lapse OCT imaging showed that the convective flows were initiated right after the SSY solution droplet was loaded on the substrate. Substantial Marangoni flows were visualized in time-lapse OCT images, which were established due to the increased concentration of SSY particles near the pinned contact line during the evaporation, leading to increase in local surface tension gradient along with the droplet interface. In the next stage, the formation of the nematic phase (N) pushed the isotropic (I)-nematic phase boundary to the center. Finally, the isotropic phase diminished at the droplet center. A volcano-shaped final deposition pattern is observed for SSY solution droplets.

Recently, Edwards et al. utilized OCT to investigate the flow patterns of the drying binary liquid droplets (**Figure 1E**) [29]. Low concentration of solvents, i.e., ethanol or n-butanol, was diluted in the water solution to form binary liquid droplets. To monitor the drying binary liquid droplets, a 1300 nm OCT system was employed, with an axial resolution of 5.5 μm and a lateral resolution of 13 μm.

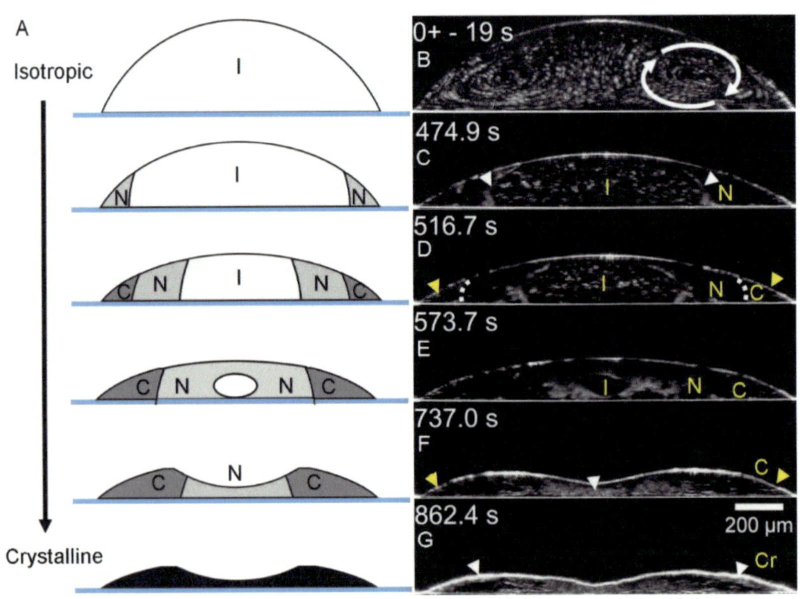

Figure 2.
Drying process of SSY solution droplet imaged by ultrahigh-resolution optical coherence microscopy (UHR-OCM). A schematic illustration of phase behaviors of SSY solution droplet at different drying stages was shown in (A). (B–G) UHR-OCM images of the SSY solution droplet at different drying time points. White spots: polystyrene particles as tracers. Cr: crystalline. C: columnar. N: nematic. I: isotropic. White dashed line in (D): columnar-nematic boundary. White arrows in (C): nematic-isotropic boundary. Yellow arrows: columnar line structures. White arrow in (G). Crystallized state. Image reproduced from Ref. [3].

A scan range of 2.5 mm was set, which was sufficient to cover the whole droplet. Different from the previous three studies, they investigated the flow pattern under different tilting conditions, including the sessile droplet mode with 0° tilting and pendant droplet mode with 180° tilting. In the tilted configuration, axisymmetric toroidal flow patterns were observed by OCT in the middle of drying processes. In the sessile droplet, a convective flow pattern in the same direction as expected by Marangoni driven flow was observed. However, in the pendant droplet, the flow direction was opposite to the Marangoni flow direction, suggesting that Marangoni flows might not dominate in the evaporating binary liquid droplets. They suggested that the convective flows were driven by the density of the liquid and gravity. In the case of the water-ethanol droplet with the preferential evaporation of ethanol, the binary liquid close to the air-water interface was denser than the bulk. Thus, surficial binary liquids would fall along the air-water interface from top to edge in sessile droplets and from edge to bottom in the pendant droplets, driving convective flows in different patterns. Also, they monitored the full drying process of the binary liquid droplets with OCT, showing a three-stage drying process, which are chaotic flows, convective flows, and outward "coffee ring" flows.

3. Drying process of colloidal latex droplets

A latex (or emulsion polymer) is a stable colloidal system with polymer particles suspending in an aqueous solution. The size of polymer particles ranges from a few nanometers to a few hundred, and the sedimentation can be neglected [31, 33]. A latex is usually synthesized by the emulsion polymerization procedure [31, 34]. Thanks to their ability of film formation, latexes can be used in application including the binder in waterborne paints [35], waterborne pressure-sensitive adhesives [36], inkjet printing [37, 38], sunscreen [39], paper coating [40, 41], drug tablet coating [42, 43], carpet backing [44], and evaporative lithography [45–47].

Drying process of latex is inhomogeneous. That is, latex particles distribute nonuniformly, spatially, and temporally. As the latex continues drying, the inhomogeneous distribution can lead to the formation of drying defects. The drying inhomogeneity can occur in both horizontal and vertical directions. In the horizontal direction, the evaporation rate is faster on the droplet edge than in the center. Further, the "coffee ring" flows drive the particles from the center to the edge, resulting in a final "coffee ring" deposition pattern, an uneven coating surface [32, 48, 49]. In the vertical direction, particles aggregate on the top surface, forming a "skin layer" that inhibits the drying process [33, 50–54]. Furthermore, if the glass-transition temperature (T_g) of particles is above room temperature, particles are stiff, and cracks would be seen in the latex [54–58]. To create uniform evaporation, it is important to understand and characterize the drying latex with different T_g, particle size and surfactant concentrations.

Using OCT to characterize the drying process of the waterborne latex droplets has been an active research effort, with an emphasis on observing drying inhomogeneity. In recent studies by Huang et al. [54, 59], the waterborne latex droplets contained polystyrene particles with different sizes, initial solid contents, and surfactant solid contents (L latex: particle size, ~125 nm; initial solid content, 40.11 wt%; surfactant solid content, 0.2 wt%. S latex: particle size, ~53 nm; initial solid content, 33.43 wt%; surfactant solid content, 1.4 wt%). To monitor the drying process of the latex droplets, the integrated OCT-gravimetry-video platform [30] was employed. The spectral-domain OCT system in the integrated platform had a central wavelength of 1320 nm and a spectral bandwidth of 110 nm. Axial and lateral resolutions of the system were 6.8 μm and 14 μm, respectively. The camera's scan rate was set to 20.7 kHz. The maximum imaging depth and lateral scan range were 2.2 mm and 5 mm. In the experiments, latex droplets with a volume of ~5 μL were loaded on a cleaned glass slide. Experiments were conducted in the ambient condition. The total drying time for both L and S latex droplets were ~13–15 min. Time-lapse, M-mode (repeated frames) OCT imaging was initiated after ~80 s from the loading of latex droplets, with a time interval of ~10 s.

Figure 3 showed time-lapse OCT imaging of drying L and S latex droplets. In **Figure 3A** and **B**, a domain boundary was clearly observed inside the L latex droplet with distinct scattering properties. The outer layer had lower scattering intensity, and inner layer had higher scattering intensity. OCT speckle contrast analysis [30, 54] was further carried out on the same data. In the outer layer, the higher speckle contrast indicated that motions of particles in these regions were restricted. The dark center inside the L latex droplet inferred that the particles were active in Brownian motions. With these analyses, we confirmed our observation that particles' packing occurred from the droplet edges and propagated inward, similar to the drying latex coat in the Petri dish [30]. The packing of particles was also observed in S latex droplet, in both OCT structural images (**Figure 3D–F**) and speckle images (**Figure 3J** and **K**). Particles began packing on the air/latex interface. In **Figure 3I**, we observed that the horizontal packing process was much faster than the vertical packing process in L latex, which was attributed to faster evaporation rate at the pinned contact line at droplet edge than the apex of the droplet. In S latex droplet, the vertical packing process was delayed as compared to the horizontal packing process. At ~410 s, we could barely see a thin layer in **Figure 3F**, indicating the existence of a vertical packed layer. The vertical layer was clearly visible until ~500 s. As a comparison, the vertical packed layer was clearly visible in the L latex droplet at ~300 s. This can be explained by the difference of diffusivity ($D = k_B T/6\pi\eta R$) [60]. Larger particles have less diffusivity than small particles, and thus small particles tend to counteract the drying and impede the packing process on the top. The particle droplets were fully packed at ~410 s for L latex droplet and ~630 s for S latex droplet.

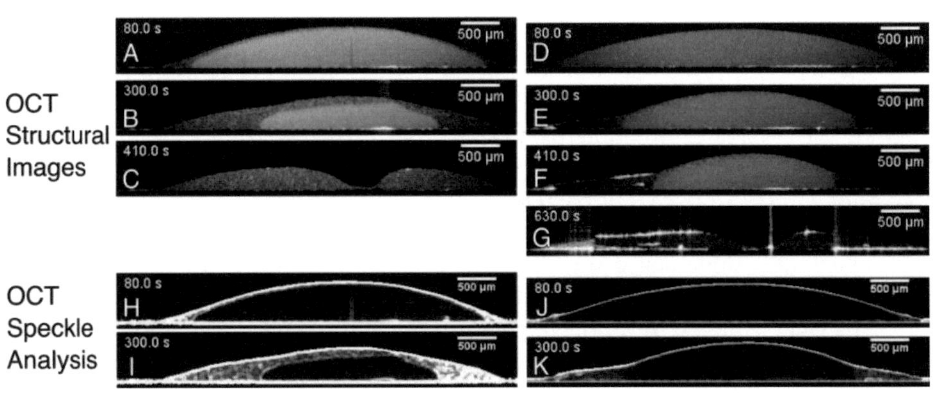

Figure 3.
Drying progression of latex droplets showing the horizontal and vertical packing process. ~5 μL latex droplets containing larger (L latex; particle diameter, ~125 nm) and smaller (S latex; particle diameter, ~53 nm) polystyrene particles were loaded on the glass slide, with initial solid contents of 40.11 and 33.43 wt%, respectively. OCT structural images of L (A–C) and S (D–G) latex droplets clearly showed domain boundaries between surficial packed region close to air-latex interface and inner suspension regions with different scattering properties. OCT speckle contrast analysis (H–K) further confirmed the inhomogeneous particles' packing process for L and S latex droplets, with different particles' mobilities in the packed and suspension regions. Image cited from Refs. [54, 59].

To sum up, the utilization of OCT was demonstrated to reveal the drying inhomogeneity in waterborne polystyrene latex droplets. The time-lapse OCT results showed that both L and S latex droplets exhibit particles packing processes in horizontal and vertical directions, but with a difference of packing speed between them. This high T_g particle can be seen as a model system to illustrate the effect of particle compaction only without the complication of particle consolidation or deformation. Further research using low T_g latex particles can potentially shine light on the full process of film formation, including particle compaction and consolidation.

4. Drying process of latex coats

OCT can also be employed to monitor the drying process of the latex coat. As shown in **Figure 4A**, OCT scans a small area on a large uniform latex coat in a Petri dish. Given the nondestructive-imaging capability, OCT can perform the cross-sectional imaging of the drying process of latex coat that simulates the real paint/coating applications and provide a characterization of internal structures, different drying behaviors, and drying inhomogeneity along the vertical direction of the latex coat. Previously, Lawman and Liang [61] and Saccon et al. [62] have reported using OCT to monitor the drying process of varnish, in which average refractive index, surface roughness, and thickness have been characterized. Huang et al. furthers the application of OCT on investigating the drying phenomena of latex coat, including the cross-sectional imaging and particles' dynamics analysis [30, 54].

Combining OCT with gravimetric and video measurements can fully characterize the drying process of polystyrene latex coat [30]. Among these modalities, time-lapse OCT imaging can show the local microstructures, 1D vertical drying process, and drying inhomogeneity from the cross-sectional view of the drying latex. **Figure 4** shows a representative OCT imaging of a latex coat containing polystyrene particles (L latex). The composition of L latex coat with particle size ~125 nm is the same as the colloidal latex droplet in Section 3. The L latex coat was loaded in an 8.5-cm-diameter Petri dish, with an initial thickness of ~1 mm. The visual appearance of a drying latex coat is presented in **Figure 4A**, with solid content (k_s) of 42.9 wt% at ~30 min. **Figure 4B** shows time-lapse OCT characterization result of the full drying process of the L latex coat. On the top of the time-lapse OCT intensity profile, four drying stages were labeled based on OCT observations of drying phenomena, including the packing process, consolidation, stress relaxation, and final drying stage. In **Figure 4B**, a linear decrease of the thickness of the L latex coat in the packing stage and the detachment of the latex film bottom in the stress relaxation stage were shown. During the stress relaxation stage, the scattering intensities change, suggesting the rearrangement of particles to release the internal stress accompanied by the infiltration of air. In the final drying stage, the latex coat remains uniform without any significant changes of the scattering intensity or thickness.

Figure 4C shows a demonstration of OCT to visualize particles' packing process in L latex coat. In zoomed-in time-lapse OCT intensity profile within the first ~180 min (**Figure 4C**, top), the separation of packed and suspension layers can be seen based on the scattering light intensity variation. The packed/suspension domain boundary is visible and highlighted by a dotted curve. Next, a speckle contrast analysis [30] identifies particles' dynamics in these two domains. In time-lapse OCT speckle image (**Figure 4C**, bottom), the particles' dynamics in these two domains can be clearly distinguished. In the upper domain, the high speckle contrast (K_s) indicates that particle

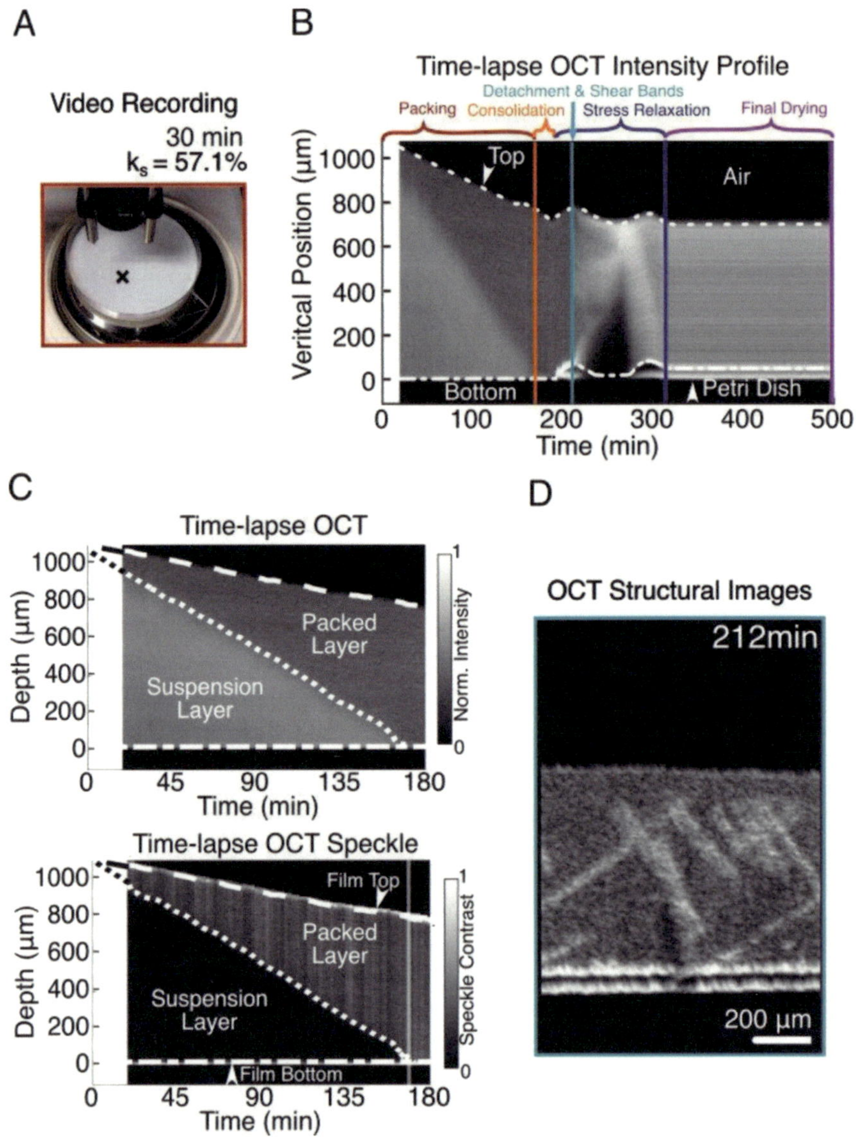

Figure 4.
OCT characterization of drying process of a polystyrene latex in Petri dish. (A) Visual appearance of the drying latex coat at ~30 min. (B) Time-lapse OCT intensity profile showing the drying process of the latex coat. (C) Visualization of particles' packing process of latex coat in time-lapse OCT intensity profile (top) and in time-lapse OCT speckle profile (bottom). (D) Visualization of latex detachment and formation of shear-band structures in 2D OCT structural image. Image reproduced from Ref. [30].

movement is restricted in the packed layer. Conversely, the low K_s value in the lower domain infers that these particles are freely moving in the suspension layer. Based on speckle contrast, the packed/suspension domain boundary can be easily identified and plotted. After the domain boundary is identified, the packed layer thickness changes can be derived from the time-lapse OCT speckle image.

Formation of shear bands is observed in the polystyrene latex coat, shown in 2D OCT structural image in **Figure 4D**. At ~212 min, the shear-band structure starts to form, indicated by the bright crosses inside the latex. The observed shear-band structure in cross-sectional OCT image is similar to the shear-band structure investigated by Yang et al. [63] and Kiatkirakajorn and Goehring [64]. The shear band is postulated to be attributed to the dislocation of packed latex particles due to the internal compressive stress along the vertical direction. These dislocations allow the air to infiltrate, resulting in a high refractive index mismatch between the air and the latex polymer or water in the dislocation, making the shear bands visible under OCT [30].

5. Discussions

In the previous sections, the feasibility of OCT is shown to characterize the drying processes of colloidal droplets and latex coats. OCT can provide cross-sectional views to observe the internal structures of the colloidal droplets and latex coats. Novel observations of shear-band structure and particles' packing process in the latex coat are shown. Based on the cross-sectional images, quantitative analyses can be conducted on drying droplets and latex coats, including contact angle of the droplet and the thickness of the latex coat. With the high-speed imaging capability, OCT can track different types of fluid flows with the aid of micron-sized tracer particles, especially the convective fluid flow. By doing the speckle contrast analysis, the packing process and the coffee ring phenomenon can be accurately imaged. The nondestructive nature of OCT enables monitoring of a full drying process to reveal the time-dependent changes, such as the phase changes of liquid crystal droplets and the consolidations of latex particles. Taken all these advantages, OCT can play an important role in fundamental studies of drying colloidal droplets and latex coats.

In OCT images of the drying colloidal materials, the fluid flows have been visualized by tracking the trajectories of micron-sized tracers. The choice of the tracers in colloidal droplets depends on the optical transparency as well as the OCT system resolutions in both axial and lateral resolutions. Empirically, the colloids with a low initial concentration of suspended materials are transparent, and the colloids with a high concentration of suspended materials (e.g., colloidal latex droplets) are semi-transparent or opaque. In order to distinguish the tracers, the light scattering properties (e.g., refractive index) of the tracers should be significantly different from the bulk colloidal droplet. Based on light scattering properties, polystyrene, gold, and titanium oxide (TiO_2) particles can be potential candidates for tracers. The size of tracers should be close to the OCT system resolution to resolve individual tracers.

For latex coats, the field of view (FOV) for a standard OCT system (a few millimeter square) only covers a small area of the latex coat. Expanding FOV for OCT systems may enable the observation of time-dependent horizontal drying inhomogeneity, such as drying front propagation. One simple approach to expand FOV is to use an objective with a lower numerical aperture, with a trade-off of the lateral resolution. An alternative approach is to utilize parallel beams to image the latex at different spots. A parallel-imaging OCT system with a space-division multiplexing technique was demonstrated by Huang et al. previously to perform wide-field imaging with simultaneous eight-channel illumination, covering an area of 18.0×14.3 mm^2 [65].

We should note that OCT measures the optical path length instead of absolute distance in the vertical direction. To derive the latex thickness, the measured optical path length value is divided by the predetermined refractive index values, assuming the refractive index remains relatively constant in space and time. Temporal and spatial variations of the refractive index, which is not considered in our experiments, may result in quantification errors in time-lapse thickness measurements.

6. Conclusions and future perspectives

In this chapter, recent progresses have been summarized on utilizing OCT for investigation of drying processes of the colloidal droplets and latex coats.

In colloidal droplets, high-speed OCT imaging can show different types of flows with the help of micron-sized tracer particles, involving the radial coffee ring flows, Marangoni flows, and density-driven flows. In addition, phase evolutions of liquid crystal droplets can be observed in OCT cross-sectional images. For colloidal latex droplets and latex coats, 1D or 2D particles' packing process can be visualized with speckle contrast analysis to characterize particles' dynamics, without adding any tracer particles. Long-term structural changes of drying latex can be investigated continuously, such as crack formation, detachment, and shear-band structures in the polystyrene latex. In these studies, the advantages of OCT are shown to provide the cross-sectional views of the droplets and latex coats with good depth-resolvability, deep penetration, good temporal resolution, and the capability of long-term, nondestructive characterization.

OCT has opened new opportunities to facilitate the fundamental studies of the interface and colloidal science to characterize different drying models. Other than sessile or pendant droplets, OCT can characterize levitated droplets to monitor different types of flows or particle motions. 1D confined droplet [66] or 1D confined flows in microfluidic channels can be imaged by OCT to characterize particles' dynamics. For latex coats, it is possible to explore the effects of additives (such as surfactant and water-soluble polymers) on the drying process of latex coats with the established integrated OCT system. In the future, OCT can be used as a tool for in-line evaluation of polymer coatings and paints for industrial applications.

Acknowledgements

We thank Jinyun Zou, Arjun Yodh, and Anne Bouchaudy for the helpful discussions on the colloidal droplet and latex coat experiments. The work is supported in part by Beijing Oriental Yuhong Waterproof Technology Co., Ltd., Emulsion Polymers Institute in Lehigh University, NSF DBI-1455613, and NIH R01-EB025209 grants.

Conflict of interest

The authors declare no conflict of interest.

Author details

Yongyang Huang[1], Hao Huang[2,3,4], Zhiyu Jiang[2], Lanfang Li[2], Willie Lau[5], Mohamed El-Aasser[3,6], Hsin-Chiao Daniel Ou-Yang[2] and Chao Zhou[1,7]*

1 Department of Electrical and Computer Engineering, Lehigh University, Bethlehem, PA, USA

2 Department of Physics, Lehigh University, Bethlehem, PA, USA

3 Department of Chemical and Bimolecular Engineering, Lehigh University, Bethlehem, PA, USA

4 Axalta Coating Systems, Philadelphia, PA, USA

5 Beijing Oriental Yuhong Waterproof Technology Co., Ltd, Beijing, China

6 Department of Materials Science and Engineering, Lehigh University, Bethlehem, PA, USA

7 Department of Biomedical Engineering, Washington University in St. Louis, St. Louis, MO, USA

*Address all correspondence to: chaozhou@wustl.edu

References

[1] Tarafdar S, Tarasevich YY, Dutta Choudhury M, Dutta T, Zang D. Droplet drying patterns on solid substrates: From hydrophilic to superhydrophobic contact to levitating drops. Advances in Condensed Matter Physics. 2018;**2018**:24

[2] Eral HB, Augustine DM, Duits MHG, Mugele F. Suppressing the coffee stain effect: How to control colloidal self-assembly in evaporating drops using electrowetting. Soft Matter. 2011;7:4954-4958

[3] Davidson ZS, Huang Y, Gross A, Martinez A, Still T, Zhou C, et al. Deposition and drying dynamics of liquid crystal droplets. Nature Communications. 2017;**8**:15642

[4] Jeong J, Davidson ZS, Collings PJ, Lubensky TC, Yodh AG. Chiral symmetry breaking and surface faceting in chromonic liquid crystal droplets with giant elastic anisotropy. PNAS. 2014;**111**:1742-1747

[5] Sadek C, Schuck P, Fallourd Y, Pradeau N, Le Floch-Fouéré C, Jeantet R. Drying of a single droplet to investigate process–structure–function relationships: A review. Dairy Science and Technology. 2015;**95**(6):771-794

[6] Deegan RD. Pattern formation in drying drops. Physical Review E. 2000;**61**(1):475-485

[7] Shimokawa Y, Kajiya T, Sakai K, Doi M. Measurement of the skin layer in the drying process of a polymer solution. Physical Review E. 2011;**84**(5):051803

[8] Anyfantakis M, Geng Z, Morel M, Rudiuk S, Baigl D. Modulation of the coffee-ring effect in particle/surfactant mixtures: The importance of particle-Interface interactions. Langmuir. 2015;**31**(14):4113-4120

[9] Kajiya T, Kaneko D, Doi M. Dynamical visualization of "coffee stain phenomenon" in droplets of polymer solution via fluorescent microscopy. Langmuir. 2008;**24**(21):12369-12374

[10] Gokhale SJ, Plawsky JL, Wayner PC. Spreading, evaporation, and contact line dynamics of surfactant-laden microdrops. Langmuir. 2005;**21**(18):8188-8197

[11] Kim H, Boulogne F, Um E, Jacobi I, Button E, Stone HA. Controlled uniform coating from the interplay of Marangoni flows and surface-adsorbed macromolecules. Physical Review Letters. 2016;**116**(12):124501

[12] Kaya D, Belyi VA, Muthukumar M. Pattern formation in drying droplets of polyelectrolyte and salt. Journal of Chemical Physics. 2010;**133**(11):114905

[13] Kovalchuk NM, Trybala A, Starov VM. Evaporation of sessile droplets. Current Opinion in Colloid and Interface Science. 2014;**19**(4):336-342

[14] Sefiane K. Patterns from drying drops. Advances in Colloid and Interface Science. 2014;**206**:372-381

[15] Wang JZ, Zheng ZH, Li HW, Huck WTS, Sirringhaus H. Dewetting of conducting polymer inkjet droplets on patterned surfaces. Nature Materials. 2004;**3**(3):171-176

[16] Dugas V, Broutin J, Souteyrand E. Droplet evaporation study applied to DNA chip manufacturing. Langmuir. 2005;**21**(20):9130-9136

[17] Trantum JR, Baglia ML, Eagleton ZE, Mernaugh RL, Haselton FR. Biosensor design based on Marangoni flow in an evaporating drop. Lab on a Chip. 2014;**14**(2):315-324

[18] Gulka CP, Swartz JD, Trantum JR, Davis KM, Peak CM, Denton AJ, et al. Coffee rings as low-resource diagnostics: Detection of the malaria biomarker *Plasmodium falciparum* histidine-rich protein-II using a surface-coupled ring of Ni(II)NTA gold-plated polystyrene particles. ACS Applied Materials & Interfaces. 2014;**6**(9):6257-6263

[19] Yakhno TA, Yakhno VG. Structural evolution of drying drops of biological fluids. Technical Physics. 2009;**54**(8):1219-1227

[20] Yakhno TA, Kazakov VV, Sanina OA, Sanin AG, Yakhno VG. Drops of biological fluids drying on a hard substrate: Variation of the morphology, weight, temperature, and mechanical properties. Technical Physics. 2010;**55**(7):929-935

[21] Rapis E. A change in the physical state of a nonequilibrium blood plasma protein film in patients with carcinoma. Technical Physics. 2002;**47**(4):510-512

[22] Bahmani L, Neysari M, Maleki M. The study of drying and pattern formation of whole human blood drops and the effect of thalassaemia and neonatal jaundice on the patterns. Colloids and Surfaces a-Physicochemical and Engineering Aspects. 2017;**513**:66-75

[23] Kokornaczyk MO, Dinelli G, Marotti I, Benedettelli S, Nani D, Betti L. Self-organized crystallization patterns from evaporating droplets of common wheat grain leakages as a potential tool for quality analysis. The Scientific World Journal. 2011;**11**:1712-1725

[24] Devlin NR, Loehr K, Harris MT. The separation of two different sized particles in an evaporating droplet. Aiche Journal. 2015;**61**(10):3547-3556

[25] Abdel-Mageed HM, Fouad SA, Teaima MH, Abdel-Aty AM, Fahmy AS, Shaker DS, et al. Optimization of nano spray drying parameters for production of α-amylase nanopowder for biotherapeutic applications using factorial design. Drying Technology. 2019:1-9. Available from: https://www.tandfonline.com/doi/full/10.1080/07373937.2019.1565576?scroll=top&needAccess=true

[26] Arpagaus C. A novel laboratory-scale spray dryer to produce nanoparticles. Drying Technology. 2012;**30**(10):1113-1121

[27] Trantum JR, Eagleton ZE, Patil CA, Tucker-Schwartz JM, Baglia ML, Skala MC, et al. Cross-sectional tracking of particle motion in evaporating drops: Flow fields and interfacial accumulation. Langmuir. 2013;**29**(21):6221-6231

[28] Manukyan S, Sauer HM, Roisman IV, Baldwin KA, Fairhurst DJ, Liang H, et al. Imaging internal flows in a drying sessile polymer dispersion drop using spectral radar optical coherence tomography (SR-OCT). Journal of Colloid and Interface Science. 2013;**395**:287-293

[29] Edwards AMJ, Atkinson PS, Cheung CS, Liang H, Fairhurst DJ, Ouali FF. Density-driven flows in evaporating binary liquid droplets. Physical Review Letters. 2018;**121**(18):184501

[30] Huang H, Huang YY, Lau W, Ou-Yang HD, Zhou C, El-Aasser MS. Integrating optical coherence tomography with gravimetric and video analysis (OCT-gravimetry-video method) for studying the drying process of polystyrene latex system. Scientific Reports. 2018;**8**:12962

[31] Lovell PA, El-Aasser MS. Emulsion Polymerization and Emulsion Polymers. 1997

[32] Deegan RD, Bakajin O, Dupont TF, Huber G, Nagel SR, Witten TA. Capillary flow as the cause

of ring stains from dried liquid drops. Nature. 1997;**389**:827-829

[33] Routh AF. Drying of thin colloidal films. Reports on Progress in Physics. 2013;**76**(4):046603

[34] Keddie J, Routh AF. Fundamentals of Latex Film Formation: Processes and Properties. Berlin: Springer Laboratory; 2010

[35] van der Kooij HM, Sprakel J. Watching paint dry; more exciting than it seems. Soft Matter. 2015;**11**(32):6353-6359

[36] Jovanovic R, Dube MA. Emulsion-based pressure-sensitive adhesives: A review. Journal of Macromolecular Science Polymer Reviews. 2004;**C44**(1):1-51

[37] Elgammal M, Schneider R, Gradzielski M. Preparation of latex nanoparticles using nanoemulsions obtained by the phase inversion composition (PIC) method and their application in textile printing. Colloids and Surfaces a-Physicochemical and Engineering Aspects. 2015;**470**:70-79

[38] Guan Y, Tawiah B, Zhang LP, Du CS, Fu SH. Preparation of UV-cured pigment/latex dispersion for textile inkjet printing. Colloids and Surfaces a-Physicochemical and Engineering Aspects. 2014;**462**:90-98

[39] Borase HP, Patil CD, Suryawanshi RK, Patil SV. Ficus carica latex-mediated synthesis of silver nanoparticles and its application as a chemophotoprotective agent. Applied Biochemistry and Biotechnology. 2013;**171**(3):676-688

[40] Najafi SMH, Tajvidi M, Bousfield DW. Production and mechanical characterization of free-standing pigmented paper coating layers with latex and starch as binder. Progress in Organic Coating. 2018;**123**:138-145

[41] Wu YM, Duan HD, Yu YQ, Zhang CG. Preparation and performance in paper coating of silicone-modified styrene-butyl acrylate copolymer latex. Journal of Applied Polymer Science. 2001;**79**(2):333-336

[42] Nakagami H, Keshikawa T, Matsumura M, Tsukamoto H. Application of aqueous suspensions and latex dispersions of water-insoluble polymers for tablet and granule coatings. Chemical & Pharmaceutical Bulletin. 1991;**39**(7):1837-1842

[43] Ladika M, Kalantar TH, Shao H, Dean SL, Harris JK, Sheskey PJ, et al. Polyampholyte acrylic latexes for tablet coating applications. Journal of Applied Polymer Science. 2014;**131**(7):12

[44] Brentin RP. Latex coating systems for carpet backing. Journal of Coated Fabrics. 1982;**12**(2):82-91

[45] Schulz M, Romermann H, Pohl K, Chindawong C, Johannsmann D. Latex films with In-plane composition gradients caused by lateral drying. Soft Materials. 2015;**13**(3):138-143

[46] Kolegov KS. Simulation of patterned glass film formation in the evaporating colloidal liquid under IR heating. Microgravity Science and Technology. 2018;**30**(1-2):113-120

[47] Georgiadis A, Muhamad FN, Utgenannt A, Keddie JL. Aesthetically textured, hard latex coatings by fast IR-assisted evaporative lithography. Progress in Organic Coating. 2013;**76**(12):1786-1791

[48] Routh AF, Russel WB. Horizontal drying fronts during solvent evaporation from latex films. Aiche Journal. 1998;**44**(9):2088-2098

[49] Divry V, Gromer A, Nassar M, Lambour C, Collin D, Holl Y. Drying mechanisms in plasticized latex films: Role of horizontal drying fronts. The

Journal of Physical Chemistry. B. 2016;**120**(27):6791-6802

[50] Routh AF, Zimmerman WB. Distribution of particles during solvent evaporation from films. Chemical Engineering Science. 2004;**59**(14):2961-2968

[51] Narita T, Hebraud P, Lequeux F. Effects of the rate of evaporation and film thickness on nonuniform drying of film-forming concentrated colloidal suspensions. European Physical Journal E. 2005;**17**(1):69-76

[52] Koenig AM, Weerakkody TG, KeddieJL,JohannsmannD.Heterogeneous drying of colloidal polymer films: Dependence on added salt. Langmuir. 2008;**24**(14):7580-7589

[53] Carter FT, Kowalczyk RM, Millichamp I, Chainey M, Keddie JL. Correlating particle deformation with water concentration profiles during latex film formation: Reasons that softer latex films take longer to dry. Langmuir. 2014;**30**(32):9672-9681

[54] Huang H. A novel dielectrophoresis potential spectroscopy for colloidal nanoparticles. Bethlehem: Lehigh University; Theses and Dissertations. 2018:4289

[55] Tirumkudulu MS, Russel WB. Cracking in drying latex films. Langmuir. 2005;**21**(11):4938-4948

[56] Tirumkudulu MS, Russel WB. Role of capillary stresses in film formation. Langmuir. 2004;**20**(7):2947-2961

[57] Birk-Braun N, Yunus K, Rees EJ, Schabel W, Routh AF. Generation of strength in a drying film: How fracture toughness depends on dispersion properties. Physical Review E. 2017;**95**(2):022610

[58] Francis LF, McCormick AV, Vaessen DM, Payne JA. Development and measurement of stress in polymer coatings. Journal of Materials Science. 2002;**37**(22):4717-4731

[59] Huang Y. Development of High-Speed Optical Coherence Tomography for Time-Lapse Non-Destructive Characterization of Samples. Bethlehem: Lehigh University; Theses and Dissertations. 2019:5564

[60] Routh AF, Russel WB. Deformation mechanisms during latex film formation: Experimental evidence. Industrial and Engineering Chemistry Research. 2001;**40**(20):4302-4308

[61] Lawman S, Liang HD. High precision dynamic multi-interface profilometry with optical coherence tomography. Applied Optics. 2011;**50**(32):6039-6048

[62] Saccon FAM, de Oliveira FMDR, Ribas MO, Zambianchi P, Muller M, Fabris JL. Kinetics of varnish long-term drying process monitored by a heterogeneous optical sensor system. Measurement Science and Technology. 2013;**24**(9):094013

[63] Yang B, Sharp JS, Smith MI. Shear banding in drying films of colloidal nanoparticles. ACS Nano. 2015;**9**(4):4077-4084

[64] Kiatkirakajorn P-C, Goehring L. Formation of shear bands in drying colloidal dispersions. Physical Review Letters. 2015;**115**(8):088302

[65] Huang Y, Badar M, Nitkowski A, Weinroth A, Tansu N, Zhou C. Wide-field high-speed space-division multiplexing optical coherence tomography using an integrated photonic device. Biomedical Optics Express. 2017;**8**(8):3856-3867

[66] Bouchaudy A, Salmon J-B. Drying-induced stresses before solidification in colloidal dispersions: In situ measurements. Soft Matter. 2019;**15**(13):2768-2781

Chapter 8

OCT for Examination of Cultural Heritage Objects

Piotr Targowski, Magdalena Kowalska, Marcin Sylwestrzak and Magdalena Iwanicka

Abstract

Optical coherence tomography (OCT) was first time reported as a tool for examination of cultural heritage objects in 2004. It is mainly used for the examination of subsurface structure of easel paintings (such as varnishes and glazes) and has also been successfully used for inspection of other types of artworks, provided that they contain layers that are permeable to the probing light. This chapter discusses the last applications of OCT in this area with an emphasis on synergy with some other noninvasive techniques such as large-scale X-ray fluorescence (XRF) scanning and reflective Fourier transform infrared (FTIR) spectroscopy. After this part, there is a detailed description of the high-resolution OCT instrument developed by the authors specifically for the study of works of art. Next, two examples are given for the structural examination of works of art: in the former, the subsurface layers of an easel painting are presented, and in the latter, the painting on reverse of the glass is examined, when the inspection must be carried out through the glass. Finally, an application for the assessment of chemical varnish removal from an easel panel painting is discussed in details.

Keywords: artwork, painting, varnish, heritage science, cleaning of paintings

1. Introduction

Optical coherence tomography is a quite natural choice for examination of objects of art. This is because it is possible to make it portable and it is contactless and noninvasive. As for the former, it is significant that the distance to an examined object is usually relatively high—in a range of centimeters rather than millimeters. It is especially important for the fragile pieces of art examined in situ, with the portable instrument mounted on a tripod or similar provisional stand. As for the latter, the intensity of the probing radiation and low energy of infrared protons ensure lack of physicochemical damage to any material: let us consider an instrument with the power of the probing beam at the object of 1 mW. If the Fourier domain fast OCT system is considered, it is reasonable to assume that the 15-mm wide B-scan is acquired in less than 0.1 s and composed of, let us say, 3000 A-scans. It leads to the scanning speed over 150 mm/s and 33 μs/A-scan. If the spot diameter will be about 12 μm, what is a reasonable measure of the lateral resolution, the fluence of the OCT beam can be estimated at 30 mJ/cm^2. This quantity is far below any damage thresholds for long pulses of infrared radiation, but a certain care in the case of work with photosensitive objects must be adopted [1].

2. An overview of recent applications of OCT in heritage science

The noninvasiveness of OCT encouraged heritage science researches to seek its applications to examination of objects of art. This subject is present in the literature since 2003 when first reports appeared [2–5]. The major fields of application have been, from the very beginning, examination of subsurface, semitransparent layers of easel paintings, glazes at faience and ceramics, historic glass, jade, and occasionally some other materials. The main restriction is in limited transparency to the probing light of materials constituting those objects. This issue has been a subject of detailed examination by Liang et al. [6] in search for the optimum wavelength window for examination. As expected the mid-IR band, about 2 μm, was found as more suitable from the point of view of transparency of pigments and dyes usually present in easel paintings. However, OCT instruments (at least commonly available) working in this range exhibit lower axial resolution due to the known trade-off between resolution and the central wavelength of operation. Nevertheless, the bibliography [7] of the subject counts now over 130 positions which is a negligible amount in comparison to those dedicated to medical applications, but constituting a little community of researchers permanently devoted to this subject. The subject has been reviewed twice already [8, 9], so for the rest of this chapter, only some of the last reports (published after 2014) will be commented on. For the complete bibliography, the reader is directed to the aforementioned website.

As for the further development of the technique, a paper by Cheung et al. [10] may serve as a mature example of the system working in 800 nm band and taking a full advantage of the axial resolution available in the near-infrared range. Due to the careful design and very broad light source (with FWHM of ~200 nm after shaping to Gaussian-like function), the axial resolution of 1.8 μm (1.2 μm in varnish-like medium of $n_R = 1.5$) is reported. The system permits resolving varnish layers as thick as 2–6 μm what is sufficient for most of examinations of easel paintings. Unfortunately, this was not a case with the second system [11] reported by the same group: the use of the supercontinuum source with the bandwidth of 220 nm working in the 2 μm range resulted in very good transparency of many layers, not permeable in 900 nm band, but the measured axial resolution was 13 μm in air which may not be sufficient for some applications.

Numerical post-processing of the results is one of the most decisive factors of the advance applications, going beyond simple feasibility studies. Usually, the solutions are adopted from the medical applications, and some concepts developed in our laboratory will be mentioned in the next sections. Here we would like to refer the reader to the paper by Callewaert et al. [12] devoted to the important subject of segmentation of thin layers, leading to, e.g., at least semiautomatic determination of thickness of superficial structures like varnish layers. This may pave the way to more efficient, or even fully automatic, monitoring of some restoration processes, even with the use of laser ablation of unwanted layers.

The area of applications of OCT to heritage objects is still expanding. Over the last few years, reports on the novel use of OCT for rock vulnerability assessment [13] and protective coatings on metal [14] have emerged. Moreover, recently the combined use of OCT with some other techniques, to take advantage of the synergy effect, has been reported. It is a known deficiency of the OCT that it does not provide information on chemical composition of structures visualized. To overcome this problem, it may be combined with other techniques, sensitive to the composition of the object, but lacking an in-depth resolution. The most promising solution seems to be linking OCT with X-ray fluorescence (XRF), especially with its macro-imaging modality MA-XRF. This combination has been presented in application for examination of the seventeenth-century Dutch still life painting [15, 16] —namely, to localize within the structure of paint layers the presence of zinc white and to clearly associate it with overpaintings present due to former restoration attempt and not with the original structures. It was important for dating of the painting since zinc white had been in common use since the second quarter of the nineteenth century only [17] and therefore, if present in the original paint layer, would shift the attribution and dating of the painting toward modern times. As for the application of MA-XRF and OCT to other kinds of artworks, in case of the late sixteenth-century illuminated manuscript (the gradual) [18], the initials made with cobalt glass pigment (smalt) and text written with iron-gall ink were in focus of the research and were well visualized by means of both OCT and MA-XRF. Another imaging technique used in combination with OCT is multispectral infrared reflectography. It permits inspection of the object in different wavelength ranges, beyond the OCT probing light [19–21]. The concept of expanding the examination window even further lies at the foundations of the complementary use of OCT and terahertz time domain imaging presented by Koch Dandolo et al. [22]. This preliminary test has shown that, as expected, the THz imaging overcomes the major limitation of OCT caused by the limited permeability of paint layer and allows imaging down to the canvas support. However, there are two other factors, which restrict the use of the THz imaging alone: the limited contrast sometimes impairs the ability to differentiate various structures (e.g., varnish versus paint) and the axial resolution, about tenfold lower than in the case of OCT. All this especially predesignate these techniques for complementary use.

The capabilities of the common use of OCT and nonlinear microscopy (NLM) were investigated by Liang et al. [1]. It was shown that NLM may in some cases provide a better contrast to differentiate between varnishes. On the other hand, this technique must be applied with care, to avoid light-induced damage of especially light-sensitive objects.

The holistic approach, including VIS-NIR multispectral imaging, high-spectral resolution VIS-NIR spectroscopy with fiber-optic reflectance spectroscopy (FORS), micro-Raman spectroscopy, XRF spectroscopy, and OCT imaging, was used by Kogou et al. [23] for examination of Chinese watercolors. In this case OCT could have been utilized as a complementary tool only, since watercolors do not exhibit permeable layers possible to examine with OCT. Therefore, it was applied in order to investigate a structure of supporting papers. Even though these materials are also essentially not permeable, some conclusions were possible to be drawn.

Properties of coatings on wood were intensively investigated lately with optical coherence microscopy exclusively [24], with OCT and hyperspectral imaging [25], as well as with OCT and synchrotron radiation micro-computed tomography (Sr-micro-CT) [26]. In this last case, the subject of examination was five large fragments removed during past restorations from historic string instruments produced by famous Italian historical violin makers: Jacobus Stainer, Gasparo da Salò, Giovanni Paolo Maggini, and Lorenzo Guadagnini. This research may be considered as feasibility study for planned examination of historic violins and showed a significant complementarity of both tomographic approaches. SR-micro-CT has better axial resolution enabling imaging some thin varnish layers, but in some cases, OCT provides better contrast permitting identification of a thin preparation layer spread over the wood, not seen with micro-CT.

The development of the OCT technique as a whole over the last 15 years as well as its applications to the examination of cultural heritage objects has permitted resolving of specific conservation issues, posted by art restorers and objects' curators.

Preceding the restoration campaign, performed in Opificio delle Pietre Dure in Florence, the unfinished masterpiece by Leonardo da Vinci "Adoration of the Magi" was extensively examined with various techniques. Among others, OCT was used [27] to determine the condition and structure of secondary varnish layers covering the painting.

OCT examination is especially effective when it is used in combination with other techniques. One of the first holistic attempts was examination of the "Bessarion Reliquary" at Opificio delle Pietre Dure in 2012 [28]. The research was directed toward the understanding of the character of specific damages of the varnish covering the painted parts of the object and tracing of remains of the historical restoration commissioned by cardinal Bessarion in the fifteenth century. Apart from OCT also point-wise XRF examination was used. Additionally, on samples collected from the object, gas chromatography with mass spectroscopy detection (GC-MS), pyrolysis gas chromatography with mass spectroscopy detection (PGC-MS), scanning electron microscopy with energy dispersive spectroscopy (SEM-EDS), and Fourier transform infrared (FTIR) spectroscopy techniques were used. Consequently, the results had been used for planning of the restoration campaign (already completed).

The combination of SEM and SEM-EDS with OCT was used by Yang et al. [29] to investigate optical properties and structure of Chinese Song Jun glaze on porcelain, especially the presence of copper and quartz additives.

During the same examination campaign as for the "Adoration of the Magi," another painting by Leonardo and studio was examined: "The Lansdowne Virgin of the Yarnwinder" ("Madonna dei Fusi"). The results of its investigation with OCT, multispectral scanning, and more common techniques, X-radiography and UV-excited fluorescence, were published recently [21]. Data post-processing with the innovative generation of scattering maps from a given depth under the surface clearly revealed the shape and in-depth location of vast overpaintings, not seen clearly with other, conventional techniques. Probably, these interventions were performed to hide damage created during two transfers of paint layer: from wood panel to canvas and from canvas to composite rigid support. The evidence of the first transfer canvas, not existing presently, was found with OCT examination as well in a form of an imprint in the paint layer. Another multi-instrumental examination campaign, the comprehensive study of Amsterdam version of "Sunflowers" by Vincent van Gogh at Van Gogh Museum in Amsterdam, was performed in 2016 as a transnational access MOLAB activity [30] of H2020 IPERION CH project. The research was aimed at documentation of a state of preservation, possible threats related to exposition, and resolving the history of restorations of this masterpiece. The goal of the OCT research was to determine the number and thickness of varnish layers and the stratigraphy within the restored areas. OCT was also used for examination of deterioration phenomena typical for this painting: local darkening of varnish in the recesses of the brush strokes, migration of the paint layer into the varnish, as well as presence and potential development of lead soap formations [31].

Apart from these reports, devoted to examination of the structure and state of preservation of artworks, OCT has been also used to monitor or assess some restoration treatments. One of the early reports, from 2011, was about application of OCT to real-time monitoring of consolidation of paint layer in reverse painting on glass (*Hinterglasmalerei*) objects [32].

Due to transparency of varnish layer, OCT is especially well suited to monitor its removal. The most common restoration technique used for this purpose is with the use of solvents. OCT, especially if used together with FTIR spectroscopy examination, is capable of accurate assessment of the cleaning process [33, 34]. Some tests of this kind were also performed during the aforementioned examination campaign of van Gogh's "Sunflowers" [35]. Selected aspects of OCT data processing for this purpose will be addressed in Section 5.

The use of lasers for ablation of unwanted layers (such as varnishes or mineral deposits) from paintings and other objects is a particularly delicate operation which needs precise control. Feasibility studies presented at LACONA (Lasers in the Conservation of Artworks) conferences in 2007, 2013, and 2015 [36–38] proved that it is possible to efficiently monitor such a process in real time. A specific case of removal of unwanted layers from the substrates of mural paintings was addressed by Striova et al. firstly in the case of removal of shellac varnish [39] and then of calcium oxalate layers [40], in both cases by means of laser and chemical treatment.

3. OCT instrument designed for heritage science

3.1 Opto-mechanical details

The instrument, constructed especially for examination of cultural heritage objects and being developed permanently in our laboratory, belongs to Fourier domain category with a broadband source and a spectrograph as detector. The instrument is designed as a portable one. To achieve this, a modular design composed of easy detachable parts connected with cables and optical fibers was chosen. All elements are of weight allowing easy handling by one person. Only a computer, comprising also all high power suppliers, weighs about 30 kg. This is important because objects of arts are often localized in old buildings with a limited access.

As for the operating parameters, the highest available axial resolution was chosen as the key parameter. This is because the ability to distinguish and visualize thin layers is a decisive factor for using in this area of applications. Therefore, short-infrared radiation was chosen even though the transparency of typical pigments is limited in this area and the optimum wavelength range lays around 2 μm [41]. By using portable commercial sources composed of coupled superluminescent diodes (Broadlighters: Q-870-HP or M-T-850-HP-I both from Superlum, Ireland) emitting in the range 770–970 nm (the former) and 750–960 (the latter), it was possible to achieve 3.3 μm of (measured) axial resolution in air, and 2.2 μm in medium of $n_R = 1.5$. The resolution was measured as a width of the point spread function with sidelobes suppression better than 25 dB. This last condition is very important in the case of examination of CH objects since they are often composed of thin and clearly transparent layers. In this case sidelobes may be easy misinterpreted as an evidence of a thin, additional layer just under the surface.

After the source, light passes a fiber-optic polarization controller and an optical isolator and is transmitted to the second module—a head by the optical fiber. The head comprises a fiber coupler of the interferometer and its reference and object arms. The former is built of the collimator, light attenuator, dispersion compensator (LSM02DC from Thorlabs), and a mirror mounted on a small translation stage. The latter comprises a fiber-optic polarization controller, the identical collimator and attenuator as in the reference arm, galvanometric scanners (6220H galvanometer scanners with MicroMax HP servo driver amplifier from Cambridge Technology, USA), and the telecentric lens (LCM04 from Thorlabs, F = 54 mm).

This optical setup provides lateral resolution of 12 μm with the distance to the object from the most protruding element of the lens equal to 43 mm and maximum scanning area of 17×17 mm^2. Alternatively the head can work with the LCM02 lens (F = 18 mm) that improves the lateral resolution to 6.5×6.5 mm^2 but for the price of significantly smaller distance to the object, 7 mm, and scanning area, 5×5 mm^2. A significant distance to the object is important for CH applications because it increases the safety of operation—objects sometimes have an irregular shape, and manipulating the head very close to the surface may pose a danger of direct contact with the artwork.

The head is equipped with two HR CCD cameras with USB-2 interface for precise documentation of the location of the OCT scanning and precise step motor-driven translators to position the head in plane and to control the distance to the object.

Finally, the light reflected in the reference arm and returning from the object undergoes the interference in the abovementioned fiber coupler and is transferred by the optical fiber to the third module—a spectrograph. Initially, the laboratory-built device with volume-phase holographic transmission grating optimized for 850 nm and with 1200 grooves/mm from Wasatch Photonics was used. Lately, it has been replaced by the complete Cobra CS800-840/180 spectrometer from Wasatch Photonics in order to improve mechanical stability and roll-off performance. Both spectrographs use 2048 pixel linear CCD cameras with fast 4-tap CameraLink interface.

3.2 Data processing and software

In the case of Fourier domain OCT instruments utilizing spectrometers with linear cameras to collect interference spectra, the result is the vector of numbers of a length equal to the number of pixels of the camera (usually 2048). This data vector requires a set of numerical procedures to be converted into one axial line of the tomogram (A-scan). Over the time, by accumulating the common experience, a standard set of numerical procedures required to obtain a high-quality A-scan has been established. It comprises a sequence of six steps: background (BG) subtraction (labelled as 1 in Figure 1), λ-k remapping (2), numerical dispersion compensation (3), spectral shaping (4), fast Fourier transformation (FFT - 5), and finally displaying in logarithmic scale (6). No one of these steps may be omitted if the best available quality of results is expected. The influence of each of the abovementioned procedures on the cross-sectional image (B-scan) is presented in **Figure 1**: a completely illegible tomogram obtained only by executing the Fourier transform on recorded spectra and displayed in logarithmic scale is presented in **Figure 1a**. The next panels show the effect of the background (BG) subtraction procedure (**Figure 1b**) and remapping of the interference spectrum from the wavelength λ (or just a camera) domain to the wavenumber k domain (**Figure 1c**). Numerical dispersion compensation (**Figure 1d**) is sharpening the image by compensating the residual dispersion mismatch between both arms of the interferometer. It is especially useful when data is collected from the structures located deep in the object of high dispersion. For example, it is a case when measurement is performed though thick glass sheet covering the investigated structures of the artwork (see **Figure 3** in the next chapter). Spectral shaping completes the set of procedures (**Figure 1e**), improving the sharpness of the boundaries of imaged structures by minimizing the sidelobes generated in the optical A-scan as a result of the Fourier transformation of the signal of the non-Gaussian envelope. This operation, being essentially the windowing of the data, may—however—slightly reduce the resolution of the system.

In order to obtain the resultant tomogram (B-scan) of the best quality, the whole set of numerical procedures presented above must be performed for every spectrum. Assuming earlier preparation of the necessary data common for all spectra (in particular vectors used in procedures of numerical dispersion compensation and λ-k remapping [42]), this analysis does not require complicated calculations: BG subtraction needs calculation of the difference of two vectors (2048 elements each); dispersion compensation and shaping are just multiplications by vectors—element

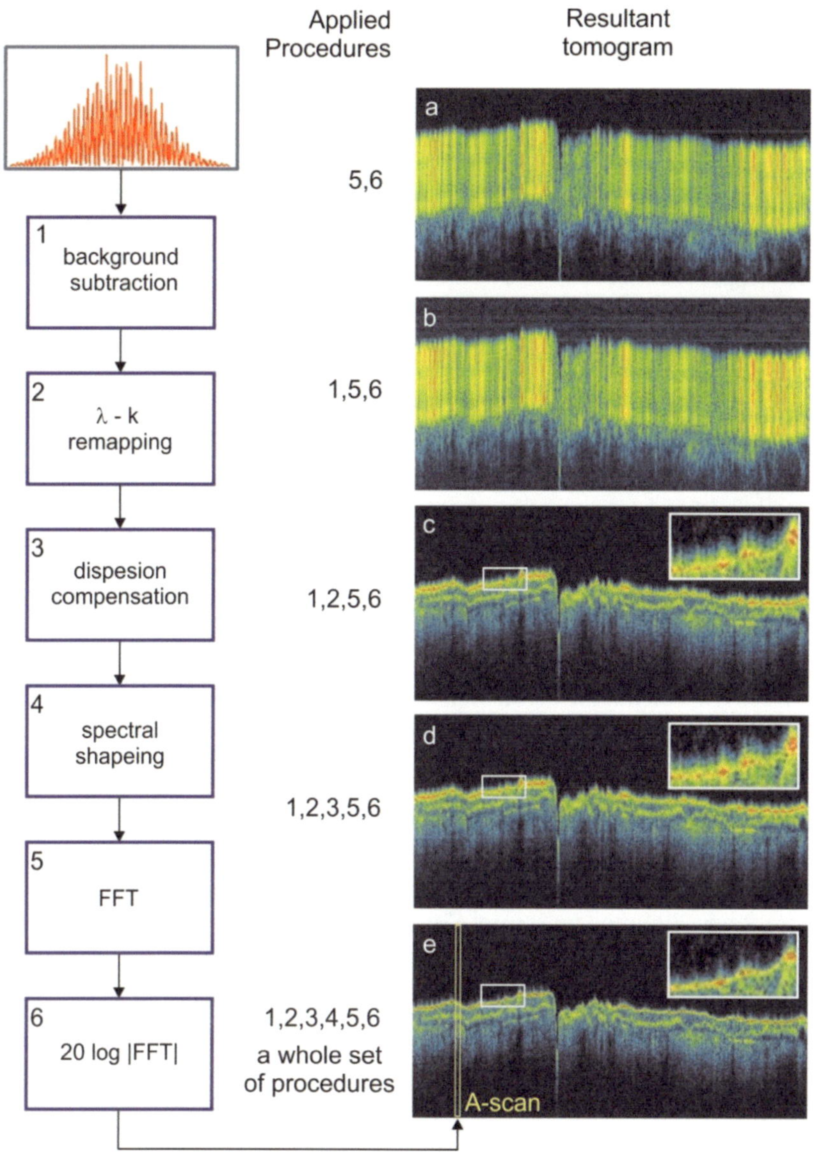

Figure 1.
The impact of individual numerical procedures on the quality of the OCT tomogram applied to a single A-scan. The procedures are labelled with numbers 1..6 and described in the text. False color scale is applied.

by element. FFT calculations are very well implemented and optimized in public domain libraries and do not take much time. λ-k remapping needs interpolation of the data (fourfold in our system, performed in Fourier domain, two FFT required) and mapping onto 2048 elements vector by a linear interpolation. The complexity of this problem arises with the number of A-scans, which must be processed—in the case of 3D volume analysis, this number typically exceeds 300,000.

Graphic processor units (GPU) are systems initially designed for fast rendering of graphic data by parallel calculations, but at present the area of applications is much broader. Often the visualization is not the major task but the massively parallel processing of any data. Obviously not all algorithms are appropriate for implementation and run fast on the GPU. However, if there is a possibility of parallelization of certain numerical tasks, the efficiency of calculations increases with the amount of data processed. In this context, the analysis of spectral optical

	Task	Time [ms] Intel Core i7-7700K GTX 1080		Time [ms] Intel Core i7-960 GTX 580	
GPU	Transfer data to the GPU	5.77	15.27	7.45	24.36
	Processing	6.86		13.20	
	Transfer of the results to the host	2.64		3.71	
CPU	Processing	940		1200	
CPU/GPU time		62		49	

Table 1.
The time profit of using GPU processing over using the main processor (CPU) measured for two systems: modern, with Intel Core i7-7700K and actually installed in our tomograph, with Intel Core i7-960.

tomography data is well suited for parallelization using the GPUs—calculations performed on successive spectra are independent of each other, and the complexity is primarily related to the amount of data being processed.

To benefit from this advantage of GPU, we have developed and successfully implemented our own software for parallel processing of the OCT data [43, 44]. It is worthwhile to note that in the case of this implementation, more time is needed to transfer the raw data to the GPU memory, and, after processing, the results back to the RAM memory of the workstation through the PCIe bus (**Table 1**). Nevertheless the profit of fast calculations is so high that the implementation of GPU is fully justified.

The efficiency of this data processing depends on the utilized hardware. On a modern workstation equipped with an Intel Core i7-7700K 4.2 GHz processor, 32 GB RAM memory, and NVIDIA GTX 1080 graphics processor, the computation acceleration is over 60 times when the process includes data transferring and almost 140 times for data processing only. The workstation actually used with our OCT configuration is equipped with Intel Core i7-960 3.2 GHz processors, 12 GB RAM, and NVIDIA GTX 580 GPU. The acceleration obtained with our software on this machine is slightly lower and amounts to 49 and 90 times (with and without data transfer). The details are presented in **Table 1**.

4. Examples of structural images of artwork

As a first example of application, OCT scans from examination of the early nineteenth-century painting on canvas "Portrait of Sir John Wylie" by F. Franck are presented (**Figure 2**). They are shown (as all tomograms in this chapter) with false color scale: structures strongly reflecting/scattering of the probing light are displayed in warm colors (red to yellow), whereas structures scattering/reflecting moderately or weekly in cold colors, from green to blue. Areas fully transparent or not accessible to the probing light are shown as black. The tomograms are corrected for refraction in the materials penetrated by the probing beam. Since most varnishes and binders have refractive indices from the range of 1.48 to 1.53 [45, 46], it is reasonable to assume an average value of 1.5 for recalculation of axial distances from optical to geometrical ones: for thin layers of varnishes possible systematic error will be below the axial resolution. The correction can be done either by the appropriate redrawing of the tomogram by application of a ray-tracing procedure taking into account ray refraction at the air-varnish boundary or by a simplified method. This approach, used in the examples presented herein, is acceptable for flat structures and is performed just by shortening of all vertical distances below the surface by a factor of 1.5. Alternatively, for uncorrected tomograms all axial optical

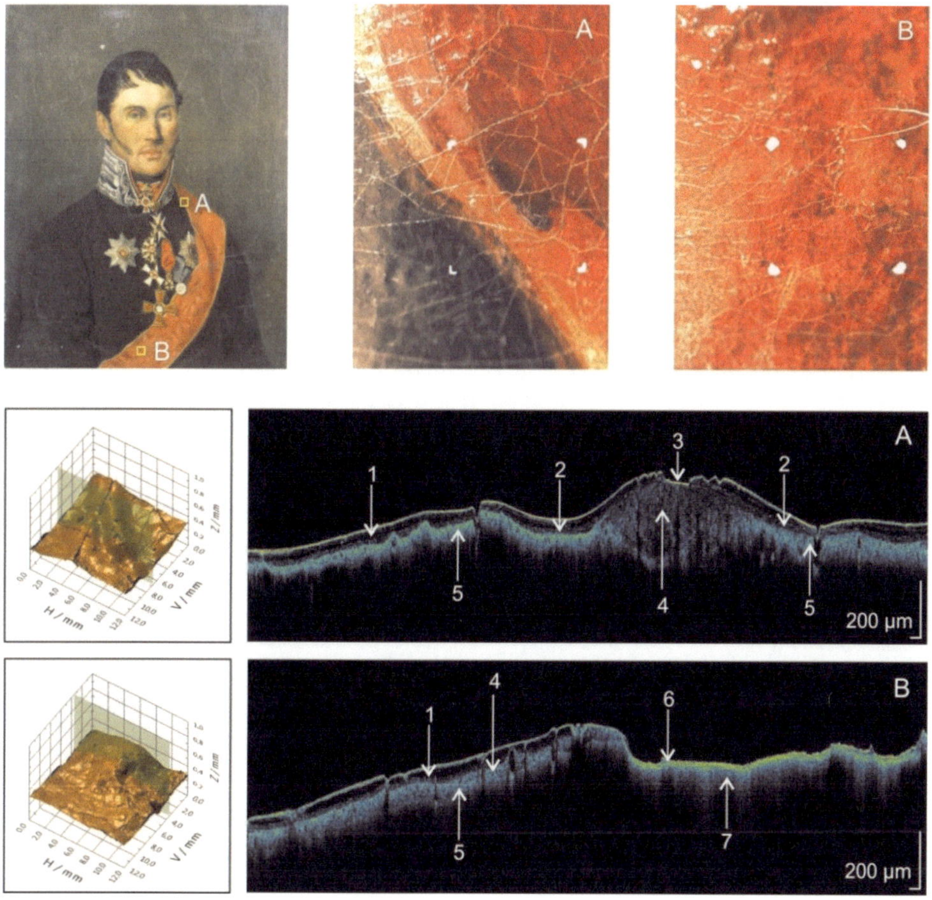

Figure 2.
OCT cross section over subsurface structures of an early nineteenth-century oil painting on canvas "Portrait of Sir John Wylie" by F. Franck. Tomograms (A, original structure of paint layers; B, restored structure at right) are corrected for refraction; thus, scale bars show geometrical distances, and false color intensity scale is used. Upper panels, from the left: a photo of the paintings with two spots (12 × 12 mm^2) of examination marked as A and B, images from the OCT annotation camera with areas of OCT scanning clearly marked. Below: exemplary tomograms and surface profiles rendered for OCT data with exact locations of OCT scans marked green. Tomograms and surface profiles are vertically stretched for better readability. Structures resolved: 1, three layers of varnish; 2, glaze layer; 3, loss in the varnish layer; 4, semitransparent paint layer; 5, opaque paint layer; 6, retouching covered by thin varnish layer; 7, putty.

distances below the surface and measured with scale appropriate for air above the object's surface (and thus for the surface topography) must be divided by 1.5. Another typical feature of tomograms used for this application is that they are vertically stretched for better readability. It is acceptable since the axial resolution is usually significantly higher.

In **Figure 2**, two examples are shown: (A) with original structure of paint layer and some secondary varnish layers and (B) area with original structure (left) and after considerable restoration (right) which included filing a paint loss with putty and reconstruction of the paint layer. Under surface of the painting, three varnish layers are seen and below, locally, a glaze layer. The last visible structure is always an opaque paint layer. If its absorption of OCT probing light is not extremely high, some multiscattering events within this layer occur, and fading tails of the signal are visible below the surface of paint (marked 5 and 7 in **Figure 2**).

Since OCT utilizes light to probe the object, it is possible to examine structures normally not accessible for inspection. As for the application to objects of art, it is often the case of reverse paintings on glass (*la peinture sur verre inversé*, *Hinterglasmalerei*) technique popular in Europe, especially favored since the middle of the eighteenth century. Later, in the nineteenth century, it has become popular in folk art, especially in Central Europe. The picture is painted on glass and intended to be viewed through it. Glass serves both as the support and the protection of the paint, which makes this technique suitable for items designed for continuous use, such as decorations of craftwork, miniatures, devotional items, etc. The major disadvantage inherent for this painting technique is an overtime decrease of adhesion of the paint layer to glass, causing delamination. Ironically, the protecting glass complicates significantly the conservation treatments since the paint layer is not accessible from the front and very often also from the back.

OCT provides in this case a convenient method for examination of the state of preservation of the object. The application to folk art was reported already [47]; here it is demonstrated for the miniature from the collection of the National Museum in Krakow, Poland (**Figure 3**). The tomogram was collected through the

Figure 3.
Through glass OCT examination of the miniature from the collection of the National Museum in Krakow (MNK III-min 933), 5.2 × 6.3 cm², reverse paintings on glass. Tomogram is not corrected for refraction; scale bars represent 200 µm in both directions. White arrow points to the early delamination of paint. Photos: Karol Kowalik, Photographic Laboratory of National Museum in Krakow, PL. Bottom right. An IR reflectogram generated from OCT data by integration over A-scans with exact localization of the tomogram.

covering glass, and thus, its upper surface is not visible since it lies beyond the imaging range of the tomograph. Therefore, the first visible line from the top is a bottom surface of glass. Below, partially attached to the support, the paint layer is evident. The gap between glass and paint may be directly measured at the tomogram and in this case varies 0.19–0.26 mm for one flake and reaches even 0.5 mm for the other one. Additionally, some traces of further delamination (marked by white arrow) are discernible. It may be interpreted as an evidence of a progressing process of destruction.

5. Application for varnish removal assessment

Varnish removal is one of the most often performed restoration treatments. Despite the fact that according to the contemporary approach to conservation/restoration of artworks, interventions should be as minimal as possible, and removal of past varnishes is a commonly accepted practice. Most common reasons for this include yellowing of varnish changing the esthetic perception of the colors of the underlying paint as well as loss of varnish transparency due to its blanching, cracking, or delaminations. The action must be taken with caution so as not to damage the paint layer underneath. In particular, if most common chemical removal is planned, a proper solvent and means of use (through controlled swabbing or gel application) must be determined.

The ability of OCT to visualize varnish layers makes it a convenient tool to monitor a varnish removal with chemical treatment and/or laser ablation qualitatively [36, 37] and quantitatively [33, 39, 40]. An application of OCT for quantitative assessment of varnish removal by swabbing is experimentally challenging because it is difficult to avoid micro-displacements of the object between sequential measurements caused by the contact with a cotton swab. The amount of material removed in one step of swabbing (or other cleaning process, e.g., laser ablation) is measured by subtraction of two surface profiles, obtained before and after treatment. However, the result will be reliable only if the position of the object in 3D space before and after is the same with a micrometer precision. As it was mentioned earlier, it is usually not possible to mechanically maintain the position of the object with the required accuracy (especially in case of paintings on canvas). Therefore, one of OCT data set must be numerically shifted in all three dimensions to achieve desired correlation of surface profiles. If the cleaning spot is significantly smaller than the area covered by OCT 3D scan, the solution is quite simple: the requested shifts may be obtained by correlating the non-treated edges of the scanned area which—in this case—will be exactly the same [33]. If, however, for any reason, the surface of the whole scanned area was altered, another procedure must be applied. In order to determine the amount of the removed material, only surface profiles obtained from OCT data are needed. However, for the proper correlation of data cubes, the inner structure—obviously not altered by the treatment—must be used. In this case it is the surface of the opaque paint layer, well visible at the tomograms. To use it as a reference, however, the tomograms must be corrected for refraction: a thickness of varnish above is by definition different, and thus, the refraction deformation is different as well, and thus, correlation of images would be systematically wrong without such a correction.

In **Figure 4** the entire procedure is illustrated on the example of a multistep test of secondary varnish removal from a panel painting. In this case, the solvent treatment by means of swabbing was chosen. In every step a cotton swab with solvent was rolled over the surface of the painting once. The aim of the test was to determine a safe amount of rolls which will not affect a paint layer and assay the homogeneity of the treatment. Before the test and after each of 13 cleaning steps, a

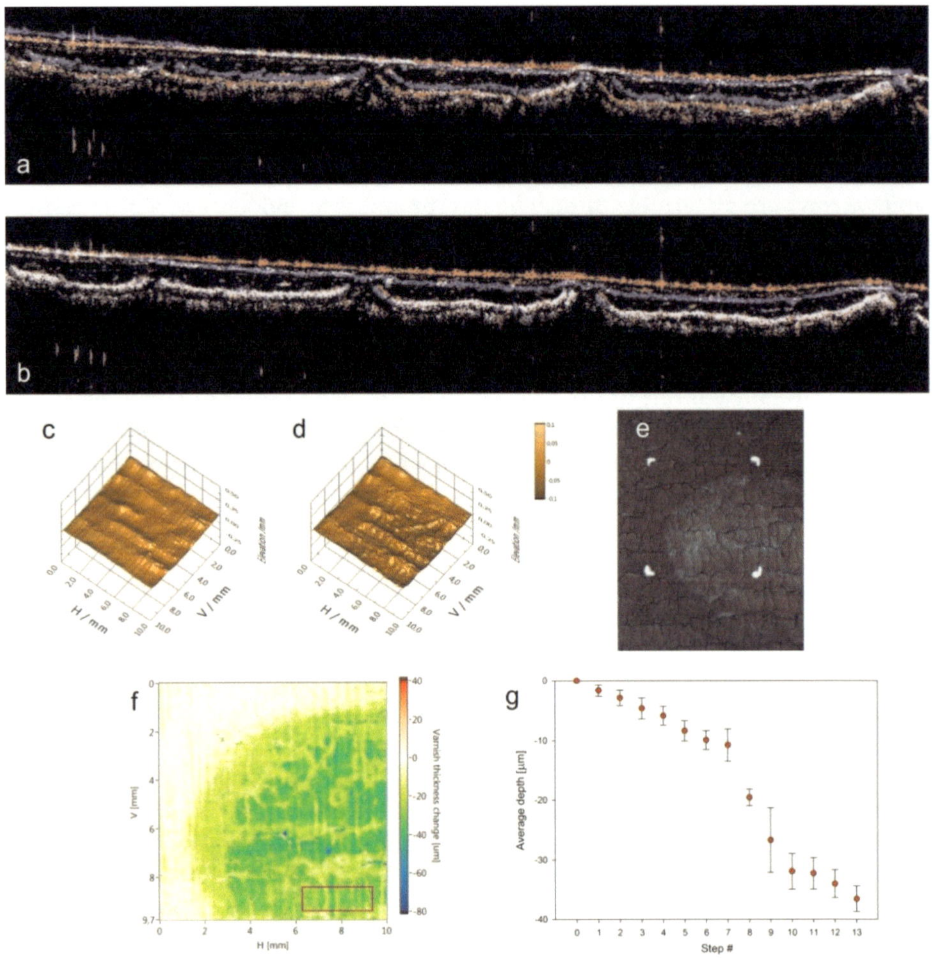

Figure 4.
Varnish removal monitoring by OCT; (a) superimposed OCT tomograms collected before (red) and after (blue) the 13th step of cleaning process, white dots represent pixels common in both tomograms, data not correlated, both tomograms corrected for refraction; (b) same but after correlation of the paint layer ($\Delta X = -30$, $\Delta Z = -13$ pixels); (c) surface profile before cleaning; (d) surface profile after cleaning; (e) photo from the OCT annotation camera taken after cleaning; (f) map of the varnish deficit after cleaning, red, area for averaging of varnish deficit; (g) average varnish thickness removed in consecutive steps.

3D OCT scan was performed over an area of 10×10 mm^2 by collection of 100 tomograms composed of 3000 A-scans each. Exemplary tomograms, chosen only for presentation in this account, are shown for all steps in Video 1 available from h ttp://repozytorium.umk.pl/handle/item/5906. Tomograms are not corrected for refraction; therefore, two scale bars are shown: for use in air and in a medium with $n_R = 1.5$. Random shifts both in X and Z directions are clearly visible. To process, at first the surface profiles were determined from each OCT 3D scan (see **Figure 4c** and **d** for two examples—before the test and after the last step, respectively). Then all OCT tomograms were corrected for refraction ($n_R = 1.5$) with simplified method as described above. As it can be seen from the exemplary tomograms (again before the test and after the last step) shown in **Figure 4a**, voxels are displaced in both X and Z directions due to unavoidable micro-displacements of the panel caused by contact with the swab. Therefore, the second tomogram had to be shifted in both directions to achieve desirable correlation of the paint layer (**Figure 4b**). It is worthwhile to note that in this particular case, displacement in Y direction, between B-scans, was determined to be smaller than the distance between adjacent scans

(0.1 mm) and thus was not taken into account. Knowing the necessary shifts, the maps of an amount varnish removed (deficit) were calculated (**Figure 4f**). For every map, the average amount was calculated over the red rectangle, and result is shown in **Figure 4g**. Since, for obvious reason, it is not possible to present in the figure results for all the steps, they can be viewed as Video 2 available from http://repozytorium.umk.pl/handle/item/5905.

Inspection of results obtained with OCT from this test permits to draw certain conclusions important for planning of the prospective restoration. Firstly, the varnish removal is not homogenous: after 13 steps it is almost completely removed from most protruding areas at the raised edges of paint layer along craquelure (fully covered by varnish before treatment), whereas it remains in about ½ of it its original thickness in recesses. Surprisingly, as it can be clearly seen from **Figure 4**, the mechanical action (during swabbing) has less impact in the raised areas along the craquelure than in the recesses. Apparently, the varnish is less soluble over the craquelure than in between. What is more, the analysis of **Figure 4g** leads to the conclusion that there is a critical range in the process around the eighth step, when it develops very quickly and the thickness of the varnish decreases rapidly. This is due to the phenomenon of swelling of varnish. Knowing at which point in the cleaning process this rapid leap in the varnish removal rate occurs aids the conservator-restorer to control the process with caution. In the case of the painting presented here, the varnish removal rate decreased in the last four steps of the OCT-monitored cleaning test. The reason for this is uncertain, one may hypothesize that the bottom varnish layers were less soluble since they were the oldest ones.

Such monitoring of the dynamic of varnish removal with OCT, even if performer locally, can then be utilized by the restorer to safely clean the whole painting, now without OCT assistance.

6. Conclusions

OCT has been used to study works of art for the last 15 years. During this time, a set of applications was developed, related to the study of the structure of the artwork, especially as a supporting tool for preventive conservation and restoration. The ability of OCT to inspect superficial layers like varnishes, glazes, and overpaintings—just to use examination of easel paining as an example—makes it especially efficient in tracing former restorations and detecting surface-related damages such as cracks, delaminations, lead soap formations, etc. In many cases, especially for most valuable artworks as well as the ones in a good state of preservation without visible losses, traditional method of investigation of the structure of an object of art by taking samples of the material is not permitted. Due to its noninvasiveness, OCT is in this case the only technique capable of visualization of the subsurface structures of works of art with desirable resolution and contrast. The OCT examination is also fast and possible to be carried out in a place where the objects are stored or displayed. This last remark is, as it is clear from the experience of the authors, very important, because in the focus of all the curators of collections is the safety of objects and their preservation for future generations. There is no doubt that optical coherence tomography contributes to this goal.

Acknowledgements

This research has been cofinanced by H2020 IPERION CH project (contract number: 654028) and conducted with the partial use of the research infrastructure of the Interdisciplinary Centre for Modern Technology of Nicolaus Copernicus University in Toruń, Poland, and financed by the Regional Operational Programme for Kujawsko-Pomorskie Voivodeship (Project No. RPKP.05.04.00-04-001/10). Results shown in **Figure 3** are obtained in cooperation with the National Museum in Krakow in the framework of the offer of the Polish Research Consortium for Heritage Science E-RIHS.PL (project leader: Zofia Maniakowska-Jazownik). A valuable contribution of Ms. Hélène Dubois from KIK-IRPA, Brussels, to conducting a cleaning test is gratefully acknowledged.

Conflict of interest

The authors declare no conflict of interest.

Author details

Piotr Targowski[1]*, Magdalena Kowalska[1], Marcin Sylwestrzak[1] and Magdalena Iwanicka[2]

1 Faculty of Physics, Astronomy and Informatics, Nicolaus Copernicus University, Toruń, Poland

2 Faculty of Fine Arts, Nicolaus Copernicus University, Toruń, Poland

*Address all correspondence to: ptarg@fizyka.umk.pl

References

[1] Liang H, Mari M, Cheung CS, Kogou S, Johnson P, Filippidis G. Optical coherence tomography and non-linear microscopy for paintings—A study of the complementary capabilities and laser degradation effects. Optics Express. 2017;**25**(16):19640-19653. DOI: 10.1364/OE.25.019640

[2] Targowski P, Rouba B, Wojtkowski M, Kowalczyk A. Application of optical coherence tomography to non-destructive examination of museum objects. Studies in Conservation. 2004; **49**(2):107-114. DOI: 10.1179/sic.2004.49.2.107

[3] Yang ML, Lu CW, Hsu IJ, Yang CC. The use of optical coherence tomography for monitoring the subsurface morphologies of archaic jades. Archaeometry. 2004;**46**(2): 171-182. DOI: 10.1111/j.1475-4754.2004.00151.x

[4] Arecchi FT, Bellini M, Corsi C, Fontana R, Materazzi M, Pezzati L, et al. Optical coherence tomography for painting diagnostics. Proceedings of SPIE. 2005;**5857**:58570Z-1-58570Z-5. DOI: 10.1117/12.612558

[5] Liang H, Gomez Cid M, Cucu R, Dobre G, Podoleanu A, Pedro J, et al. En-face optical coherence tomography —A novel application of non-invasive imaging to art conservation. Optics Express. 2005;**13**(16):6133-6144. DOI: 10.1364/OPEX.13.006133

[6] Cheung CS, Daniel JMO, Tokurakawa M, Clarkson WA, Liang H. Optical coherence tomography in the 2-μm wavelength regime for paint and other high opacity materials. Optics Letters. 2014;**39**(22):6509-6512. DOI: 10.1364/OL.39.006509

[7] Optical Coherence Tomography for Examination of Works of Art—A Complete List of Papers on Application of OCT to Examination of Artwork [Internet]. Available from: http://www.oct4art.eu [Accessed: 08 June 2019]

[8] Targowski P, Iwanicka M. Optical coherence tomography: Its role in the non-invasive structural examination and conservation of cultural heritage objects—A review. Applied Physics A: Materials Science & Processing. 2012; **106**(2):265-277. DOI: 10.1007/s00339-011-6687-3

[9] Targowski P, Iwanicka M, Rouba BJ, Frosinini C. OCT for examination of artwork. In: Drexler W, Fujimoto G, editors. Optical Coherence Tomography Technology and Applications. Cham Heidelberg New York Dordrecht London: Springer; 2015. pp. 2473-2495

[10] Cheung CS, Spring M, Liang H. Ultra-high resolution Fourier domain optical coherence tomography for old master paintings. Optics Express. 2015; **23**(8):10145-10157. DOI: 10.1364/OE.23.010145

[11] Cheung CS, Daniel JMO, Tokurakawa M, Clarkson WA, Liang H. High resolution Fourier domain optical coherence tomography in the 2 um wavelength range using a broadband supercontinuum source. Optics Express. 2015;**23**(3):1992-2001. DOI: 10.1364/oe.23.001992

[12] Callewaert T, Dik J, Kalkman J. Segmentation of thin corrugated layers in high-resolution OCT images. Optics Express. 2017;**25**(26):32816-32828. DOI: 10.1364/oe.25.032816

[13] Bemand E, Liang H, Bencsik M. Non-invasive methods for in-situ assessing and monitoring of the vulnerability of rock-art monuments. In: Darvill T, Batarda Fernandes AP, editors. Open-Air Rock-Art Conservation and Management: State of

the Art and Future Perspectives. Abingdon: Routledge; 2014. pp. 244-258

[14] Lenz M, Mazzon C, Dillmann C, Gerhardt NC, Welp H, Prange M, et al. Spectral domain optical coherence tomography for non-destructive testing of protection coatings on metal substrates. Applied Sciences. 2017;7:364. DOI: 10.3390/app7040364

[15] Iwanicka M, Ćwikliński Ł, Targowski P. Combined use of Optical Coherence Tomography and Macro-XRF Imaging for Non-invasive Evaluation of Past Alterations in 17th c. Dutch Painting. Jerusalem. 2016. Available from: http://art2016.isas.co.il/wp-content/uploads/sites/10/2017/03/Session-I-Minerals-Pigments-Dyes-3-Iwanicka.pdf [Accessed: 23 June 2019]

[16] Iwanicka M, Sylwestrzak M, Targowski P. Optical coherence tomography (OCT) for examination of artworks. In: Bastidas DM, Cano E, editors. Advanced Characterization Techniques, Diagnostic Tools and Evaluation Methods in Heritage Science. Cham: Springer International Publishing; 2018. pp. 49-59. DOI: 10.1007/978-3-319-75316-4_4

[17] Kuhn H, White Z. In: Feller RL, editor. Artists' Pigments, a Handbook of their History and Characteristics. Vol. 1. London: Archetype; 1997. pp. 169-186

[18] Targowski P, Pronobis-Gajdzis M, Surmak A, Iwanicka M, Kaszewska EA, Sylwestrzak M. The application of macro-X-ray fluorescence and optical coherence tomography for examination of parchment manuscripts. Studies in Conservation. 2015;**60**(Supplement 1): S167-SS77. DOI: 10.1179/0039363015Z.000000000221

[19] Liang H, Burgio L, Bailey K, Lucian A, Dilley C, Bellesia S, et al. Culture and trade through the prism of technical art history: A study of Chinese export paintings. Studies in Conservation. 2014;**59**(sup1):S96-SS9. DOI: 10.1179/204705814x13975704318272

[20] Striova J, Dal Fovo A, Fontani V, Barucci M, Pampaloni E, Raffaelli M, et al. Modern acrylic paints probed by optical coherence tomography and infrared reflectography. Microchemical Journal. 2018;**138**:65-71. DOI: 10.1016/j.microc.2017.12.027

[21] Targowski P, Iwanicka M, Sylwestrzak M, Frosinini C, Striova J, Fontana R. Using optical coherence tomography to reveal the hidden history of "*The Lansdowne Virgin of the Yarnwinder*" by Leonardo da Vinci and studio. Angewandte Chemie. 2018;**57**: 7396-7400. DOI: 10.1002/anie.201713356

[22] Koch Dandolo CL, Lopez M, Fukunaga K, Ueno Y, Pillay R, Giovannacci D, et al. Toward a multimodal fusion of layered cultural object images: Complementarity of optical coherence tomography and terahertz time-domain imaging in the heritage field. Applied Optics. 2019; **58**(5):1281-1290. DOI: 10.1364/ao.58.001281

[23] Kogou S, Lucian A, Bellesia S, Burgio L, Bailey K, Brooks C, et al. A holistic multimodal approach to the non-invasive analysis of watercolour paintings. Applied Physics A: Materials Science & Processing. 2015;**121**(3): 999-1014. DOI: 10.1007/s00339-015-9425-4

[24] Gurov I, Margaryants N, Zhukova E. Evaluation of art subjects implemented in the marquetry technique by the optical coherence microscopy method. Strain. 2018;**0**(0): e12304. DOI: 10.1111/str.12304

[25] Dingemans LM, PV M, Liu P, Adam AJL, Groves RM. Optical coherence tomography complemented by hyperspectral imaging for the study of protective wood coatings. Proceedings

of SPIE. 2015;**9527**:952708. DOI: 10.1117/12.2184716

[26] Fiocco G, Rovetta T, Invernizzi C, Albano M, Malagodi M, Licchelli M, et al. A micro-tomographic insight into the coating systems of historical bowed string instruments. Coatings. 2019;**9**(2): 81. DOI: 10.3390/coatings9020081

[27] Iwanicka M, Sylwestrzak M, Szkulmowska A, Targowski P. Pre-restoration condition of superficial layers of the adoration of the magi by Leonardo da Vinci as seen by optical coherence tomography. In: Ciatti M, Frosinini C, editors. Il restauro dell'Adorazione dei Magi di Leonardo La riscoperta di un capolavoro. Florence: Edifir; 2017. pp. 287-293

[28] Iwanicka M, Lanterna G, Lalli CG, Innocenti F, Sylwestrzak M, Targowski P. On the application of optical coherence tomography as a complimentary tool in an analysis of the 13th century byzantine Bessarion reliquary. Microchemical Journal. 2016; **125**:75-84. DOI: 10.1016/j.microc.2015.11.014

[29] Yang M-L, Katz JI, Barton J, Lai W-L, Jau-Ho J. Using optical coherence tomography to examine additives in Chinese song Jun glaze. Archaeometry. 2014;**57**(5):837-855. DOI: 10.1111/arcm.12125

[30] MOLAB (Mobile LABoratory)—Transnational access to facilities within IPERION CH project or E-RIHS ERIC. Available from: http://www.iperionch.eu/molab/or http://www.e-rihs.eu/access/ [Accessed 09 June 2019]

[31] van den Berg KJ, Hendriks E, Geldof M, de Groot S, van der Werf I, Miliani C, et al. Structure and chemical composition of the surface layers in the Amsterdam Sunflowers. In: Hendriks E, Vellekoop M, editors. Van Gogh's Sunflowers Illuminated Art Meets Science. Amsterdam: Amsterdam University Press/Van Gogh Museum Amsterdam; 2019. pp. 159-173

[32] Iwanicka M, Kwiatkowska EA, Sylwestrzak M, Targowski P. Application of optical coherence tomography (OCT) for real time monitoring of consolidation of the paint layer in Hinterglasmalerei objects. Proceedings of SPIE. 2011;**8084**: 80840G. DOI: 10.1117/12.890398

[33] Iwanicka M, Moretti P, van Oudheusden S, Sylwestrzak M, Cartechini L, van den Berg KJ, et al. Complementary use of optical coherence tomography (OCT) and reflection FTIR spectroscopy for in-situ non-invasive monitoring of varnish removal from easel paintings. Microchemical Journal. 2018;**138**:7-18. DOI: 10.1016/j.microc.2017.12.016

[34] Moretti P, Iwanicka M, Melessanaki K, Dimitroulaki E, Kokkinaki O, Daugherty M, et al. Laser cleaning of paintings: In situ optimization of operative parameters through non-invasive assessment by Optical Coherence Tomography (OCT), reflection FT-IR spectroscopy and Laser Induced Fluorescence spectroscopy (LIF) measurements. Heritage Science. 2019;**7**: 44-55. DOI: 10.1186/s40494-019-0284-8

[35] Hendriks E, Geldof M, van den Berg KJ, Monico L, Miliani C, Moretti P, et al. Conservation of the Amsterdam Sunflowers: From past to future. In: Hendriks E, Vellekoop M, editors. Van Gogh's Sunflowers Illuminated Art Meets Science. Amsterdam: Amsterdam University Press/Van Gogh Museum Amsterdam; 2019. pp. 175-205

[36] Góra M, Targowski P, Kowalczyk A, Marczak J, Rycyk A. Fast spectral optical coherence tomography for monitoring of varnish ablation process. In: Castillejo M, Moreno P, Oujja M, Radvan R, Ruiz J, editors. Lasers in the Conservation of Artworks Proceedings of the International Conference Lacona VII,

Madrid, Spain, 17-21 September 2007. London: CRC Press, Taylor & Francis Group; 2008. pp. 23-27

[37] Targowski P, Marczak J, Kwiatkowska EA, Sylwestrzak M, Sarzyński A. Optical coherence tomography for high resolution real-time varnish ablation monitoring. In: Saunders D, Strlič M, Korenberg C, Luxford N, Birkhölzer K, editors. Lasers in the Conservation of Artworks IX Proceedings of the International Conference on Lasers in the Conservation of Artworks (Lacona IX), 7–10 September 2011, London, UK. London: Archetype Publications Ltd; 2013. pp. 26-31

[38] Iwanicka M, Musiela J, Łukaszewicz JW, Stoksik H, Sylwestrzak M. The potential of OCT for assessing laser assisted removal of deposits from ceramic tiles. In: Targowski P, Walczak M, Pouli P, editors. Lasers in the Conservation of Artworks XI, Proceedings of the International Conference LACONA XI, Kraków, Poland, 20–23 September 2016. Torun: NCU Press; 2017. pp. 105-114. DOI: 10.12775/3875-4.07

[39] Striova J, Salvadori B, Fontana R, Sansonetti A, Barucci M, Pampaloni E, et al. Optical and spectroscopic tools for evaluating Er:YAG laser removal of shellac varnish. Studies in Conservation. 2015;**60**(S1):S91-SS6. DOI: 10.1179/0039363015Z.000000000213

[40] Striova J, Fontana R, Barucci M, Felici A, Marconi E, Pampaloni E, et al. Optical devices provide unprecedented insights into the laser cleaning of calcium oxalate layers. Microchemical Journal. 2016;**124**:331-337. DOI: 10.1016/j.microc.2015.09.005

[41] Liang H, Lange R, Peric B, Spring M. Optimum spectral window for imaging of art with optical coherence tomography. Applied Physics B. 2013; **111**(4):589-602. DOI: 10.1007/s00340-013-5378-5

[42] Szkulmowski M, Tamborski S, Wojtkowski M. Spectrometer calibration for spectroscopic Fourier domain optical coherence tomography. Biomedical Optics Express. 2016;**7**(12): 5042-5054. DOI: 10.1364/boe.7.005042

[43] Sylwestrzak M, Szkulmowski M, Szlag D, Targowski P. Real-time imaging for spectral optical coherence tomography with massively parallel data processing. Photonics Letters of Poland. 2010;**2**(3):137-139. DOI: 10.4302/plp.2010.3.14

[44] Sylwestrzak M, Szlag D, Marchand PJ, Kumar AS, Lasser T. Massively parallel data processing for quantitative total flow imaging with optical coherence microscopy and tomography. Computer Physics Communications. 2017;**217**(Supplement C:128-137. DOI: https://doi.org/10.1016/j.cpc.2017.03.008

[45] de la Rie ER. The influence of varnishes on the appearance of paintings. Studies in Conservation. 1987; **32**:1-13. DOI: 10.2307/1506186

[46] Feller RL, Stolow N, Jones EH, editors. On Picture Varnishes and their Solvents. Revised and Enlarged Ed. Seminar on Resinous Surface Coatings; 1957. National Gallery of Art: Oberlin, Ohio; 1985

[47] Iwanicka M, Tymińska-Widmer L, Rouba B, Kwiatkowska EA, Sylwestrzak M, Targowski P. Through-glass structural examination of Hinterglasmalerei by optical coherence tomography. In: Radvan R, Asmus JF, Castillejo M, Pouli P, Nevin A, editors. Lasers in the Conservation of Artworks VIII Proceedings of the International Conference on Lasers in the Conservation of Artworks (Lacona VIII), 21-25 September 2009, Sibiu, Romania. London: CRC Press, Taylor & Francis Group; 2011. pp. 209-214

Chapter 9

Quantitative Mapping of Strains and Young Modulus Based on Phase-Sensitive OCT

Vladimir Y. Zaitsev, Lev A. Matveev,
Alexander A. Sovetsky and Alexander L. Matveyev

Abstract

In this chapter we consider mapping of local strains and tissue elasticity in optical coherence tomography (OCT) based on analysis of phase-sensitive OCT scans. Conventional structural OCT scans correspond to spatially resolved mapping of the backscattering intensity of the probing optical beam. Deeper analysis of such sequentially acquired multiple OCT scans can be used to extract additional information about motion of scatterers in the examined region. Such detailed analysis of OCT scans has already resulted in creation of OCT-based visualization of blood microcirculation, which has been implemented in several commercially available devices, especially for ophthalmic applications. Another functional extension of OCT emerging in recent years is the OCT-based elastography, i.e., mapping of local strains and elastic properties in the imaged region. Here, we describe the main principles of local strain mapping in phase-sensitive OCT with a special focus on the recently proposed efficient vector method of estimation of interframe phase-variation gradients. The initially performed mapping of local strains is then used for realization of quantitative compressional elastography, i.e., mapping of the Young modulus and obtaining stress-strain dependences for the studied samples. The discussed principles are illustrated by simulated and experimental examples of elastographic OCT-based visualization. The presented elastographic principles are rather general and can be used in a wide area of biomedical and technical applications.

Keywords: optical coherence elastography, phase-sensitive OCT, strain mapping, deformation imaging, stiffness mapping, Young modulus mapping

1. Introduction

Scans in optical coherence tomography (OCT) strongly resemble those obtained by ultrasound scanners. Conventional OCT images correspond to spatially resolved visualization of the backscattering intensity for optical waves similarly to visualization of backscattering intensity of ultrasonic waves. Deeper analysis of such images, especially applied to sequentially acquired multiple scans, opens possibilities to extract a rich additional information about motion of scatterers in the examined region. The development of such functional extensions in OCT imaging was in many aspects stimulated by the analogous trends in ultrasound.

In this context, probably the most well-known extension of ultrasound-based imaging is visualization of flows (first of all, imaging of blood flows in medical ultrasound angiography [1]). In OCT, generically similar principles in the 2000s were also used to realize OCT-based visualization of blood microcirculation. Since the typical imaging depth in OCT is ~1–2 mm and lateral field of view ~several millimeters, OCT-based angiography visualizes the microcirculation on a smaller scale in comparison with ultrasound but correspondingly with a higher resolution typical of OCT, where the typical resolution is 5–15 μm. Now the angiographic modality in OCT is implemented in several commercially available devices, especially for ophthalmology [2] but also for other medical applications, e.g., in oncology [3–5].

Another functional extension of OCT, the development of which was also inspired by analogous trends in ultrasonic imaging, is the optical coherence elastography (OCE), i.e., mapping of deformations (including local strains) and elastic properties (first of all, the Young modulus) in the imaged region. This direction in OCT development was triggered in 1998 by the seminal publication by Schmitt [6]. In that paper, Schmitt considered the possibility of transferring to OCT the ideas proposed in medical ultrasound in the very beginning of 1990s [7]. This approach was rather successfully developed in subsequent years [8] and since ~2000 was realized in several commercially available ultrasound platforms.

In OCT, however, the development of elastographic mode passed with 10–15 years delay. Despite the evident similarity between ultrasound and OCT scans, the transferring to OCT, the elastographic principles successfully realized in ultrasound appeared to be rather challenging. Sufficiently successful realizations of OCT-based elastography were demonstrated only during the last ~5 years. Similarly to OCT-based microangiography, elastography in OCT is mostly focused on biomedical applications (e.g., [9–12]), but certainly similar principles attract interest for engineering applications, e.g., for testing polymers [13].

In what follows, we briefly overview the main trends in the development of OCE, including measurement of strains and quantitative OCT-based mapping of the Young modulus with focus of the so-called compressional phase-sensitive OCE that became one of the most active directions in the development of OCE.

2. Basic principles used in OCE

The basic principles used for realization of OCE generically are rather similar to those used in ultrasound imaging for elastographic purposes. The primary goal of elastography is to estimate the shear modulus of the studied material (at least in the relative sense without absolute quantification). In biomedical applications this interest is explained by the fact that all soft biological tissues have very similar value of the bulk modulus that varies very insignificantly in various states of the tissue. In contrast, the shear modulus for the same state of the tissue may exhibit much stronger variability, up to several times and even orders of magnitude. Therefore, it is the observation of shear modulus variability which is especially interesting for studying structural changes in soft materials. The materials that are called "soft" at the intuitive level formally belong to the class of "nearly incompressible" or "water-like" materials, for which their Poisson's ratio ν is very close to the upper physically allowable limit $\nu \to 0.5$, i.e., to the value typical of liquids. The condition $\nu \to 0.5$ also means that the shear modulus G of such a material is much smaller than its bulk modulus K: $G/K \ll 1$. When a rod, made of such a material with free boundaries in lateral direction, is subjected to axial loading that causes its axial strain ε_{zz}, the soft material experiences lateral strains $\varepsilon_{xx} = \varepsilon_{yy} \approx -0.5\varepsilon_{zz}$. Consequently, the expansion in one direction is compensated by contraction in the orthogonal directions, such that the volume of the so-loaded sample remains almost invariable. This explains why materials with $\nu \rightarrow 0.5$ are called "nearly incompressible."

Another remarkable feature of such materials with $\nu \rightarrow 0.5$ is a rather specific relationship between their Young modulus G and shear modulus $E = 2(1 + \nu)G \approx 3G$. Therefore, instead of direct measurements of modulus G (via observation of propagation of shear or surface waves, which is used in wave variants of both ultrasonic elastography [14] and OCE [15]), it is possible to use reaction of the tissue under longitudinal uniaxial stress to evaluate the Young modulus $E = 3G$.

The idea to use quasistatic uniaxial stress for estimation of the Young modulus was proposed for ultrasound [7] and was transferred to OCT in paper [6]. In practice, the strain, which can be characterized as fairly close to uniform and uniaxial, is created in the vicinity of a piston pressed onto a tissue that is not stuck at the piston-tissue interface and can fairly freely slide laterally. This strain is created by compressional loading, so that the idea of measuring the Young modulus in such a configuration is called compressional (or compression) elastography.

The key point in realization of compressional OCE is, therefore, estimation of axial strains in the so-compressed material by analyzing a series of OCT scans acquired during the material compression. Comparison of such scans can be used to reconstruct axial displacements $U(z)$ of scatterers, and the local axial strains can be estimated by evaluating spatial gradients of the displacement field:

$$\varepsilon_{zz} = \frac{\partial U}{\partial z} \tag{1}$$

For tracking the displacements of scatterers, conventionally correlation principles have been discussed and fairly successfully realized in various applications, including medical ultrasound elastography [8] and engineering problems, where this processing is applied to sequences of photographic images of a deformed surface [16]. In paper [6] similar principles were supposed to be transferred to the analysis of OCT images. However, the attempts to directly transfer the correlational principles of estimating strains by tracking displacements using consequently acquired OCT scans were not very successful [17] in the sense that the correlationally reconstructed displacement fields were rather noisy. In view of this, their numerical differentiation required for reconstructing local strains did not give satisfactory results.

The reason of this was that unlike photographic images, OCT scans are characterized by a peculiar speckle structure originated in OCT scans from the interference of optical waves scattered from sub-resolution scatterers. This speckle pattern is rather sensitive to deformation of the imaged material, because straining produces relative displacements of the sub-resolution scatterers (for which a quarter-wavelength mutual displacement in the axial direction changes the character of interference of the scattered waves from constructive to destructive and vice versa). In view of this, for fairly moderate strains $\sim$ a few percent and even less, speckles in OCT images may demonstrate pronounced "boiling" and "blinking" resulting in strong decorrelation of the compared OCT scans. The intuitively attractive idea to use the correlational speckle tracking for "sufficiently small strains," for which the abovementioned decorrelation could be avoided, did not help too much. The reason is that in fact both masking distortions producing the decorrelation and the variations in the speckle structure that are used for the tracking purposes do appear simultaneously and are proportional to the same order of strain [18]. Thus, even for small strains, the decorrelation of OCT images usually does not allow for sufficiently precise correlational speckle tracking allowing for performing subsequent differentiation and finding strains via Eq. (1). The possibilities of correlational speckle tracking in OCT imaging of deformed tissues were analyzed in detail [19]. The conclusion was that by the abovementioned reasons, the correlational approach to strain reconstruction in OCT can be practically feasible for super-broadband spectrum of the illuminating source (like used in [20]), for which the deformation-induced speckle blinking/boiling is strongly suppressed. Otherwise the correlation approach is operable for only approximate strain estimations, i.e., with averaging over large portions of the entire OCT scan. By this reason in what follows, we will focus on another realization of compressional OCE, based on phase-resolved measurement. This approach has proven to be rather promising and is especially actively developed in recent years.

In OCT the phase of the backscattered signal is naturally available and can be readily used to track the displacements of scatterers using the well-known relationship between the variation Φ in the backscattered-wave phase and the axial scatterer displacement U:

$$U = \frac{\lambda_0 \Phi}{4\pi n}, \tag{2}$$

where λ_0 is the optical wavelength in vacuum and n is refractive index of the material. It can be shown that the phase of the OCT signal can be more tolerant to strain-induced decorrelation [21]. Consequently, even for "typical" OCT systems (i.e., without the need to ensure a super-broadband spectrum for reduction of deformation-induced decorrelation), phase measurements related to the displacements of scatterers can be made much more reliably than the correlational speckle tracking. This is an important advantage of phase-sensitive approaches to estimation of strains.

Thus, estimation of the axial gradient of phase variations $\Phi(z)$ makes it possible to estimate local strains via Eq. (1). It is important to point out that for pixelated OCT images, both the displacements and distances are naturally measured in pixels (the physical values of both quantities being dependent on the refractive index). Therefore, the gradients of the phase variations, also calculated in pixels, give correct values of strain without the necessity to know the refractive index, which is a positive feature of phase-sensitive OCT-based strain measurements.

Another important point is that, for unambiguous relation between the observed interframe phase variation $\Phi(z)$ and displacements $U(z)$ of scatterers in the observation point, the condition $U(z) < \lambda/4$ should be fulfilled. For larger displacements the phase wrapping occurs because of periodicity of the dependence of parameters of a wave on its phase. In view of this, the interframe displacement can be directly extracted from the interframe phase variation only with an uncertainty to the unknown integer number of wave periods. To exclude this ambiguity in estimations of the displacements, the conventional approach to realization of unambiguous estimation is to ensure sufficiently small interframe displacements that do not exceed $\pm\lambda/4$ [22].

Alternatively, in the cases when the displacement of scatterers is caused by the material straining, the displacements may gradually increase over the OCT scan in a wide range from essentially sub-wavelength (without phase wrapping) to super-wavelength values (with phase wrapping). Although multiple phase wrappings may occur over the entire imaged depth, in order to reconstruct a continuous function $U(z)$ even for super-wavelength displacements, the gradual increase in the phase variation makes it possible to apply conventional phase unwrapping procedures by adding 2π rad. at every depth where the phase variation exhibits a 2π jump. However, the unwrapping procedure is error-prone, so that because of measurement noises always present in real measurements (especially in areas with weaker scattering and reduced signal-to-noise ratio (SNR)), the unwrapped phase may be rather noisy. Consequently, numerical differentiation of such a noisy function $\Phi(z)$ (or equivalently $U(z)$) can be made with a reasonable accuracy only with averaging over large portions of the entire imaged depth.

In what follows we consider an efficient recently developed approach to strain-reconstruction phase-sensitive OCE based on local estimation of the gradient $dU(z)/dz$ without preliminary reconstruction of the displacement function in the depth ranges where the estimates of the total displacement $U(z)$ may be ambiguous because of phase wrapping. Then some examples of the application of the phase-sensitive strain visualization for mapping both fairly rapidly varying strains and very slow varying ones (with special optimizations of the measurement procedures) will be considered. Finally, we will discuss a realization of quasistatic compressional OCE with application of reference translucent layers with pre-calibrated stiffness for obtaining quantitative estimates of the Young modulus.

3. Local estimates of strains in phase-sensitive OCE using the "vector method"

Significant progress in strain mapping has been achieved in OCE in recent years due to transition to the use of phase-resolved OCT data. Quite a detailed discussion of axial strain estimation based on phase-resolved data for compared deformed and reference OCT scans was presented in [23]. In that paper, the least-square method (including the improved version with amplitude weighting) was considered to estimate local gradients of function $\Phi(z)$ using averaging over a window with a size significantly smaller than the entire image size. In [24, 25] another procedure for finding phase gradients was proposed. That approach was called "vector method," because it operates with complex-valued OCT signals (i.e., with signals characterized by amplitude and phase). Such signals can be considered as vectors in the complex plane, which explains why the method is called "vector." The estimated phase gradient is singled out at the very last stage of the signal processing.

Comparison with the least-square fitting of the $\Phi(z)$ slope (even with amplitude weighting to suppress noisy small-amplitude pixels) for the same processing window size demonstrated superior robustness of the vector method with respect to strain-induced speckle-decorrelation noise and other measurement noises [24]. One of the advantages of the vector method is due to the fact that the amplitude weighting is also intrinsically made in the vector method (since the signal amplitude determines the absolute value of the corresponding vector) and, furthermore, especially strong phase errors exceeding $\pi/2$ rad. are naturally suppressed in this method even for strong signals. The increased tolerance of the method to decorrelation noises makes this method operable under elevated interframe strains (up to $\sim 10^{-2}$). Consequently, for comparing other noises, the possibility to operate with elevated strains with larger phase gradients corresponds to effectively higher signal-to-noise ratio (SNR). This makes the vector method especially suitable for visualization of aperiodic strains (with magnitudes in the range $10^{-3} - 10^{-2}$), for which enhancement of SNR via conventional periodic averaging is not possible.

Consider the main steps of signal processing in the vector method. Let the complex-valued signal in each pixel (m,j) in the reference scan be written as $a_1(m,j) = A_1(m,j)\exp\left[i \cdot \phi_1(m,j)\right]$ and $a_2(m,j) = A_2(m,j)\exp\left[i \cdot \phi_2(m,j)\right]$ correspond to the deformed scan. In each scan the signal amplitudes $A_{1,2}$ and phases $\phi_{1,2}$ are random because of random positions and scattering strength of scatterers.

However, the strain-induced variations in the phase are caused by fairly regular displacements of scatterers and can be found by singling out the phase in the product:

$$a_2(m,j)a_1^*(m,j) \equiv b(m,j) = B(m,j)\exp\left[i \cdot \Phi(m,j)\right] \tag{3}$$

Here, the asterisk denotes complex conjugation, $B(m,j) = A_2(m,j)A_1(m,j)$ and $\Phi(m,j) \equiv \phi_2(m,j) - \phi_1(m,j)$. Phase variation $\Phi(m,j)$ for pixel (m,j) is related to the axial displacements $U(m,j)$ of the scatterers with coordinates close to pixel (j,m) via Eq. (2).

Usually the axial strain $\partial U/\partial z$ is evaluated by finding axial gradient of the discreet phase variation $\Phi(m,j)$ with averaging within a processing window $N_x \times N_z$ pixels in size (e.g., using the least-square method [23]). Alternatively, the averaging procedures can be performed with complex-valued quantities (3) considered as vectors in the complex plane. Then the phase gradients can be singled out only at the very final stage.

In [24] the vector approach was considered for laterally weakly inhomogeneous phase variations (i.e., for nearly uniaxial straining of the material along z axis). In such a case, the complex-valued quantities $b(m,j)$ (which may exhibit phase fluctuations due to various measurement noises and strain-induced decorrelation) can be laterally averaged within a chosen processing window to obtain an array $\overline{b(j)} \equiv B(j)\exp\left[i \cdot \Phi(j)\right]$ with more regular phase $\Phi(j)$:

$$\overline{b(j)} = \sum_{m=1}^{N_x} b(m,j) = \sum_{m=1}^{N_x} A_2(m,j)A_1(m,j)\exp\left\{i \cdot \left[\phi_2(m,j) - \phi_1(m,j)\right]\right\} \tag{4}$$

Here $m = 1..N_x$ is the horizontal index of the vertical columns in the processing window, and $j = 1..N_z$ is the index of the horizontal rows (see schematic **Figure 1**). For the averaged vector $\overline{b(j)}$, contributions of noisy small-amplitude pixels become significantly suppressed, such that they do not strongly distort the phases $\Phi(j)$ of the resultant vectors $\overline{b(j)}$. Furthermore, **Figure 1**, where Eq. (4) is represented as the summation of vectors, illustrates that the strongest phase errors give minimal distortions of the averaged phase $\Phi(j)$. (See the second panel in **Figure 1**, where it is shown that the individual noisy vectors $b(m,j)$ with almost opposite directions with respect to the direction of the averaged vector $\overline{b(j)}$ weakly affect the orientation of $\overline{b(j)}$.) For nonzero strain, the so-constructed phase variation $\Phi(j)$ depends on the discrete (pixelated) vertical coordinate j. In principle, similarly to [23], the vertical phase gradient then can be found by conventional least-square fitting of $\Phi(j)$ within the vertical size N_z of the processing window as was used in [26].

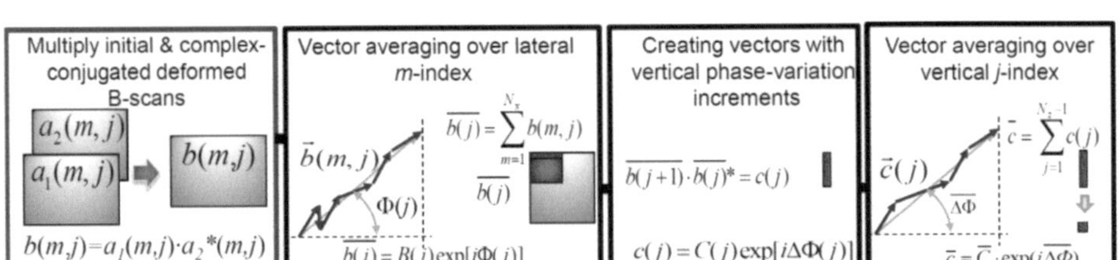

Figure 1.
Schematic of evaluation of the axial phase gradient in the vector approach for laterally nearly homogeneous strains. In the vector diagrams, the complex-valued summands b(m,j) and c(j) are shown as vectors in the complex plane. All intermediate transformations are performed with complex-valued signals, and the sought phase gradient is singled out at the last stage.

However, by analogy with the initial vector averaging, the next step of finding the vertical gradient of function $\Phi(j)$ can also be done without explicit extraction of the phases $\Phi(j)$. Namely, one can define complex-valued quantities $c(j)$ that contain vertical increments $\Delta\Phi(j) = \Phi(j+1) - \Phi(j)$ of the horizontally averaged interframe phase variations $\Phi(j)$ (see the third panel in **Figure 1**):

$$c(j) = \overline{b(j+1)\cdot b(j)^*} \equiv B(j+1)B(j)\exp\{i\cdot[\Phi(j+1) - \Phi(j)]\} \quad (5)$$

The quantity $c(j)$ can also be used in the normalized form $c_{norm}(j) = c(j)/|c(j)|^\alpha$. It was verified that good results can be obtained using normalization with $\alpha = 1$

$$c_{norm}(j) = \exp[i\cdot\Delta\Phi(j)], \quad (6)$$

which corresponds to retaining information about the phase increment only.

The complex-valued quantities $c(j)$ (or $c_{norm}(j)$) can also be considered as vectors and vertically averaged in the vector sense. If the vertical size of the processing window is smaller than the characteristic vertical scale of the strain inhomogeneity, the phase-variation increments $\Delta\Phi(j)$ are nearly identical (but in reality may be distorted by decorrelation and other noises). Vector averaging over the vertical size of the processing window then gives a complex quantity with a much more stable phase $\overline{\Delta\Phi}$:

$$\bar{c} \equiv \overline{C}\exp\left[i\cdot\overline{\Delta\Phi}\right] = \sum_{j=1}^{N_z-1} c(j)/|c(j)|^\alpha. \quad (7)$$

Summations in Eqs. (4) and (7) actually correspond to obtaining of averaged real and imaginary parts of the complex-valued signals $b(j,m)$ and $c(j)$, with subsequent singling out the resultant phase instead of direct averaging of individual phases for $b(j,m)$ and $c(j)$. As is clear from **Figure 1**, the vector summation makes the phases of the averaged vectors especially tolerant to small-amplitude erroneous vectors and signals with especially strong phase errors $\sim\pi$ rad which may occur for individual summands $b(j,m)$ or $c(j)$.

For the vertical inter-pixel distance d_p in the imaged tissue, the so-found vertical phase increment $\overline{\Delta\Phi}$ is proportional to the sought strain $\partial U/\partial z = \gamma\overline{\Delta\Phi}/d_p = \overline{\Delta\Phi}\left(\lambda_0/4\pi d_p^0\right)$ and does not depend on the refractive index. Indeed, according to Eq. (2), the coefficient $\gamma = \lambda_0/4\pi n$, and the inter-pixel distance in the material is $d_p = d_p^0/n$ (since for a material with refractive index n, the same phase of the probing optical wave is accumulated along n times smaller distance than in vacuum).

The above-considered averaging procedures are well applicable for fairly horizontal plane-parallel phase-variation isolines. However, in many case of practical interest, noticeable lateral inhomogeneities of the strain distribution may occur, so that the interframe isophase lines may be noticeably inclined within the processing window. Consequently, the directions of vectors $b(j,m)$ in Eq. (4) may strongly differ as a function of horizontal index. Thus, the straightforward lateral averaging like in Eq. (4) instead of improvement may significantly distort the visualized strain distribution.

To suppress this negative effect of averaging and retain advantages of the vector method, the following improvements can be proposed [25]. At the first stage, the complex-valued interframe signal $b(m,j)$ is averaged over small regions (having lateral and horizontal sizes $N_{small} \sim 2-3$ pixels only (see **Figure 2**)). This initial

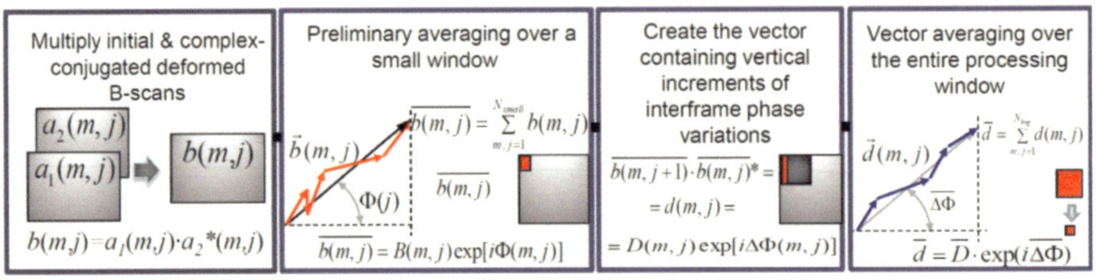

Figure 2.
Schematic of evaluation of the axial phase gradient in the vector approach adapted for laterally inhomogeneous strains with non-horizontal phase-variation isolines. All intermediate transformations are performed with complex-valued signals, and the sought phase gradient is singled out at the last stage.

averaging does not yet noticeably affect the resolution but yields less noisy interframe matrix $b(m,j) \to \tilde{b}(m,j)$ by suppressing distorting contributions of weakest amplitude and most noisy pixels. At the next step, one creates a matrix containing vertical increments of interframe phase variations (similarly to Eq. (5) but without horizontal averaging over the processing window):

$$d(m,j) = \tilde{b}(m,j+1)\tilde{b}^*(m,j) \quad (8)$$

To avoid confusion in the notations instead of array $c(j)$, here we introduced the matrix of complex-value quantities $d(m,j)$. The so-constructed matrix $d(m,j) = D(m,j)\exp[i\Phi(m,j)]$ contains the sought vertical inter-pixel increments $\Phi(m,j)$ of the interframe phase variations. The vertical increments $\Phi(m,j)$ of the interframe phase variations should already be nearly identical even if the isolines of the interframe phase variations are inclined within the processing window. Therefore, the quantities $d(m,j)$ can be efficiently averaged in the horizontal direction even for noticeably inclined interframe phase-variation isolines. Thus, the next steps are performing the vector averaging of $d(m,j)$ over the processing window with the size $N_x \times N_z$ pixels. For this procedure, the order of averaging over indices m and j is not essential, so that we first average over index m and then over j.

Note that, by analogy with the previously considered normalization of vectors $c_{norm}(j)$ containing vertical phase-variation increments, the utilization of normalized quantity $d(m,j)$ can be useful, so that with normalization the averaging over index m takes the form

$$\overline{d(j)} = \sum_{j=1}^{N_x-1} d(m,j)/|d(m,j)|^{\alpha} \quad (9)$$

Here, notation $\overline{d(j)}$ is introduced to denote averaging over the horizontal dimension of the processing window. By analogy with Eq. (7), it was verified that the simplest choice $\alpha = 1$ yields quite good results.

Similarly, averaging over the vertical index $j = 1..(N_z - 1)$ yields better results with additional normalization:

$$\overline{\overline{d}} = \sum_{j=1}^{N_z-1} \overline{d(j)}/\overline{d(j)}^{\beta}. \quad (10)$$

Here, normalization exponent $\beta = 1$ can also be recommended. Equation (10) represents a complex-valued quantity $\overline{\overline{d}} = \overline{D}\exp\left[i\overline{\Delta\Phi}\right]$ (averaged over the processing window), in which $\overline{\Delta\Phi}$ is the sought vertical inter-pixel phase variation for the current position of the processing window. Phase increment $\overline{\Delta\Phi}$ is directly analogous to the angle estimated in Eq. (7) and is also similarly linked to the sought axial strain. Geometric interpretation of Eqs. (8)–(10) is illustrated in **Figure 2**.

To illustrate the applicability of the vector approach in the most clear form, we first consider a numerically simulated example in which the reference and deformed OCT scans are simulated using model [27]. The simulation parameters for **Figure 3** correspond to a typical OCT system, for which the central wavelength is 1300 nm, spectral width 90 nm, and A-scan of 256 pixels corresponds to the imaging depth 2 mm in air. The inhomogeneity of scattering amplitudes in the simulated structural image (**Figure 3a**) are chosen similar to typical experimental scans (compared with **Figure 6**). In the simulations initially 1024 scatterers were randomly distributed over each A-scan, and then the initial positions of the scatterers were displaced according to the assumed strain distribution shown in **Figure 3b**. **Figure 3c** shows the color map for interframe phase variations corresponding to the strain distribution in **Figure 3b** showing pronounced lateral inhomogeneity and tilting of isophase lines.

The reconstructed strain obtained using the above-described vector approach is shown in **Figure 3c** and **d**. The processing window size is 16 × 16 pixels. The noisy areas in **Figure 3d** correspond to the regions of strongly inclined isophase lines, for which the straightforward lateral averaging Eq. (4) worsens the quality of elastographic mapping. **Figure 3e** shows much better strain-reconstruction quality for the modified vector method corresponding to Eqs. (8)–(10). We note that for **Figure 3**, only strain-induced decorrelation noise is taken into account. **Figure 4** illustrates the vector method tolerance to other measurement noises. The latter were simulated by adding to each pixel of the image random complex-valued numbers with Gaussian distribution in order to obtain a preselected ratio between the average intensity of the OCT image and the added noise. **Figure 4** demonstrates

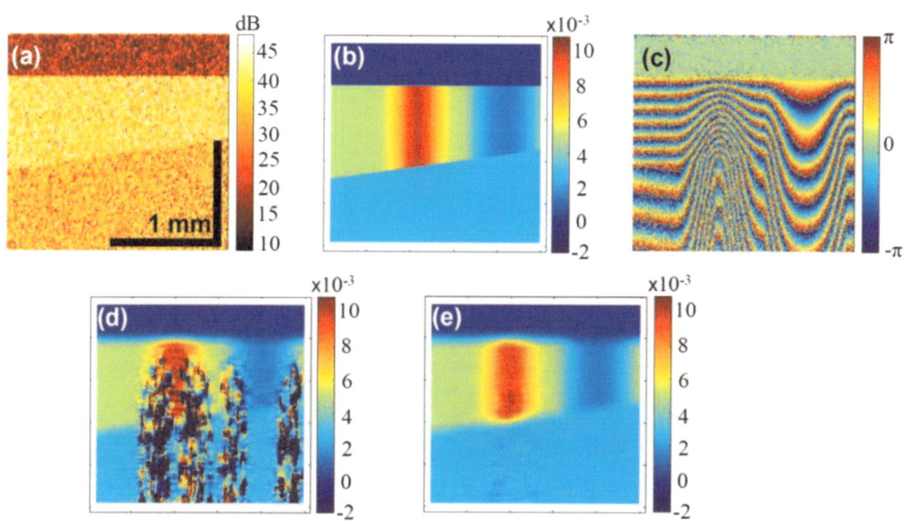

Figure 3.
Simulation of OCT-based mapping of laterally inhomogeneous strain field based on model [27]. (a) Is the simulated structural image; (b) shows the strain distribution adopted in the model; (c) color map corresponding to the interframe phase variations shown in (b); (d) is the reconstructed strain map using the vector method with straightforward horizontal averaging showing good results in laterally fairly homogeneous areas but prone to errors in the regions of inclined isophase lines; (e) is the reconstructed strain map based on the modified vector method adapted to averaging in regions of inclined isophase lines. The processing window is 16 × 16 pixels in size and the preliminary averaging area for panel (e) 2 × 2 pixels. In this example only strain-induced decorrelation noise is present.

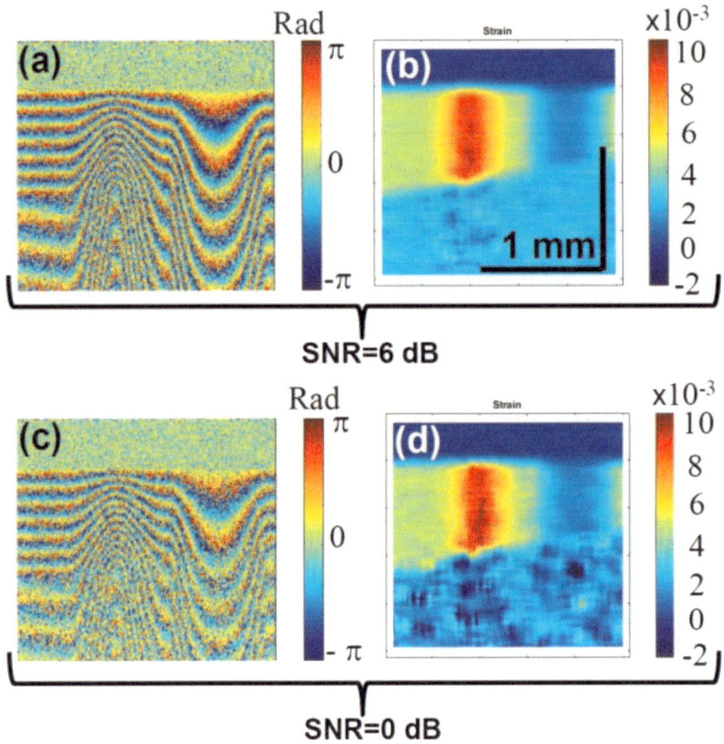

Figure 4.
Illustration of the vector method tolerance to measurement noises (other than strain-induced decorrelation noises) for the same simulated data as in ***Figure 3*** *but with additional noise (in the form of Gaussian random complex values added to each pixel). Panels (a) and (c) show the interframe phase-variation maps in the presence of noises for SNR = 6 dB and SNR = 0 dB, respectively. Panels (b) and (d) show the corresponding strain maps reconstructed using the vector method adapted for inclined phase-variation isolines.*

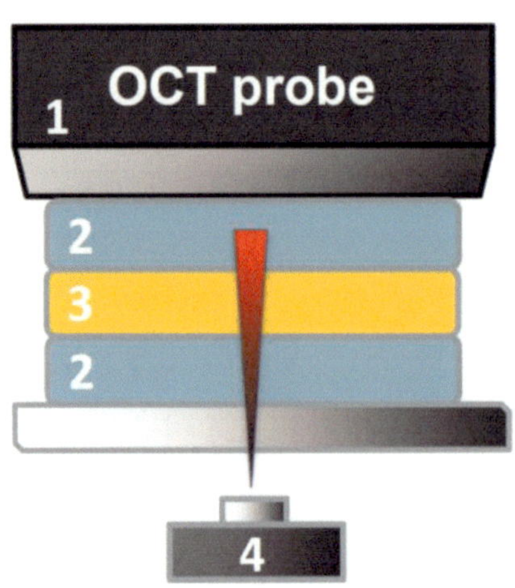

Figure 5.
Schematic of OCT imaging setup: 1 is the OCT scanner (λ = 1.3 μm); 2 is the soft silicone layer; 3 is the studied biopolymer sample (e.g., samples of cartilaginous or corneal tissues); 4 is the source of the heating infrared irradiation (erbium fiber laser operating at λ = 1.56 μm).

that the vector approach ensures fairly satisfactory reconstructions of strain maps down to average SNR∼1 (i.e., 0 dB) over the image area.

4. Examples of the vector method application for strain mapping by elastographic processing real OCT scans

To illustrate operability of the vector method for elastographic processing of real phase-sensitive OCT scans, we demonstrate some examples of mapping thermally induced interframe strains which were produced in samples of biopolymers (collagenous samples of cartilage and eye cornea) by pulse-periodic irradiation of the sample by an infrared laser operating at a wavelength of 1.56 μm that is efficiently absorbed in the water that is present in the tissue. The heating pulses had duration ∼ a few seconds and the temperature reached ∼50–60°C in the corneal tissue and ∼60–80°C in cartilaginous samples. In more detail the experimental conditions are described in papers [28–30]. The used custom-made OCT device had parameters close to those assumed for the simulated examples in **Figures 3** and **4**. A typical experimental configuration is shown in **Figure 5**. The measurements were made in contact mode such that the OCT-probe surface contacted the studied sample though an intermediate layer of translucent silicone with pre-calibrated Young modulus. Certainly, measurements in the noncontact mode when the studied samples had free boundaries were also possible. However, during the irradiation of the water-saturated biopolymers, the silicone layer played a useful auxiliary role to protect the sample from drying during the heating. Furthermore, the main destination of such a layer was to play the role of compliant sensor to estimate the pressure exerted by the deformed silicone onto the studied sample (similarly to the discussion in [31]). OCT-based monitoring of strains in the pre-calibrated reference silicone and studied sample during mechanical compression of the silicone-sample sandwich was used to obtain stress-strain curves for the studied samples and estimate their Young modulus.

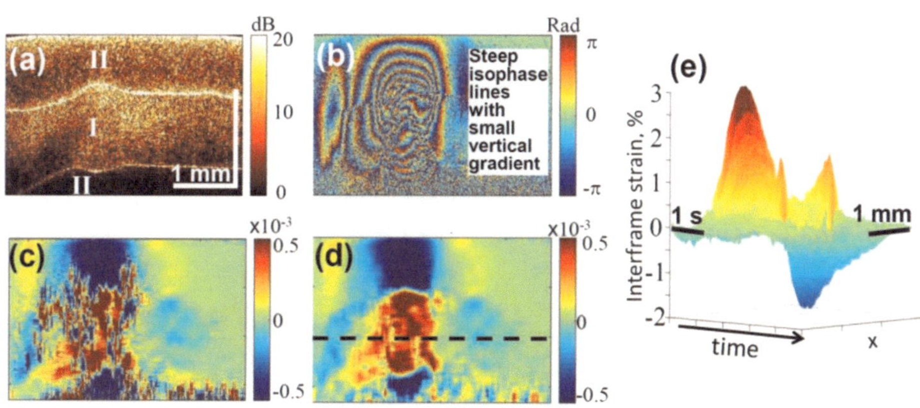

Figure 6.
Experimental demonstrations of OCT-based mapping of interframe strains. (a) is a structural OCT image of a biopolymer layer I (rabbit cornea sample) placed between translucent silicone layers (II); (b) is a typical color map of interframe phase difference, where the heated region is characterized by pronounced lateral inhomogeneity; (c) and (d) are the strain maps obtained with straightforward horizontal averaging over the processing window and using the vector method adapted for processing tilted isophase lines, respectively; (e) is a waterfall image showing the time evolution of the interframe-strain profile at the depth labeled by the dashed line in panel (d).

Figure 6 gives an example of experimentally reconstructed maps of interframe ("instantaneous") strain obtained using the experimental configuration shown in **Figure 5** for a rabbit cornea layer during one of the heating pulses. Structural image in **Figure 6a** demonstrates that the irradiated region exhibits visually appreciable local expansion after a series of heating pulses, and **Figure 6b** shows the interframe phase-variation map with pronounced inhomogeneity in the lateral direction with regions of very steep isophase lines. Similarly to simulated **Figure 3**, the strain map **Figure 6c** demonstrates the interframe-strain map found using straightforward horizontal averaging within the processing window (with sizes ~1/20 of the entire image sizes), and **Figure 6d** is obtained using the method variant adapted for processing regions of steep isophase lines. **Figure 6c** and **d** clearly demonstrates that the irradiation-induced expansion of the heated water-saturated sample causes straining with the opposite sign (contraction) of the surrounding silicone layers. **Figure 6e** for one of the heating pulses demonstrates a pseudo-3D waterfall image showing the complex spatiotemporal evolution of the interframe-strain profile corresponding to the depth marked by the dashed line in the B scans in **Figure 6d**. We emphasize that the robustness of the vector method to measurement noises makes it possible to obtain rather clear quantitative strain maps for aperiodic strain evolution when periodic averaging for enhancement of SNR is impossible.

It can be said that in the above examples, strains were "instantaneous," corresponding to the time interval between the neighboring scans (for the examples in **Figure 6**, this interval was 50 ms). However, in other biomedical and technical applications, it may be interesting to monitor cumulative strains over larger time intervals, during which the studied sample may be subjected to an external action (like laser irradiation in the above examples) or may exhibit other structural variations, e.g., chemical curing of polymers, influence of drying or impregnation by a liquid, etc.

The developed technique readily makes it possible to perform the summation of interframe strains in order to find the resultant cumulative strain. The issue of finding cumulative strains is discussed in more detail in [1] since this can be done in somewhat different ways, the result of which do not strongly differ for fairly small strains, but this difference may become appreciable for larger strains (~10%). It can be said that the choice of the method of strain cumulation depends on the particular problem. For example, straightforward summation of interframe strains on larger interval may give a biased estimate of the total change in the sample thickness but may be preferable if the difference in strain over the imaged area

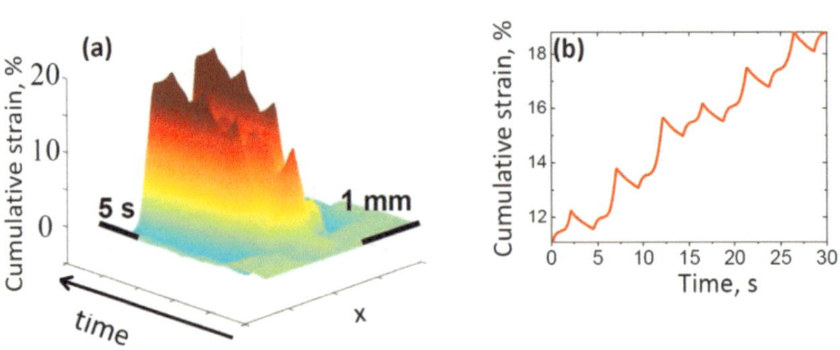

Figure 7.
*Experimental demonstrations of OCT-based mapping of cumulative strains in the same experimental configuration as for **Figure 6**. (a) is a waterfall image showing the time evolution of the cumulative-strain profile at the depth labeled by the dashed line in **Figure 6d**; (b) is a representative graph showing the cumulative strain as a function of time for a particular lateral coordinate and depth.*

should be used for estimating the differences in the elastic modulus in different image regions. Referring to [32] for details, for illustration purposes we will use simple straightforward summation of interframe strain. **Figure 7a** shows an example of the so-found evolution of lateral profile of cumulative strain (in the same experimental configuration as for evolution of interframe strains along a particular depth shown in **Figure 6e**). **Figure 7b** shows a representative graph for the time dependence of cumulative strain for a given lateral position. The profile demonstrates that during the heating, the tissue experiences internal structural changes with changes in the functional behavior upon reaching certain threshold temperatures (see peculiar inflection points in the dependence in panel **Figure 7b**). These changes are related to heating-induced active generation of pores as discussed in more detail in [30].

In examples shown in **Figures 6** and 7, we showed the results for only 2D mapping of evolving strain that evolved fairly rapidly, so that their 3D mapping via comparison between entire 3D sets of OCT data was impossible for a conventional scanning OCT system with a moderate rate of obtaining A-scans of ~20–80 kHz. Utilization of a much faster scanning system may be also problematic for sufficiently long-time monitoring because of huge data flows and total amounts of OCT data that would require special means for signal acquisition and storage, as well as supercomputing performance for real-time processing. However, there are many processes of high interest, for which strains evolve much slower (e.g., drying of a sample, gradual curing of a polymer, various osmotic phenomena [33], etc.) To monitor such slow processes with acceptable time resolution, much larger time steps ~1 sec and even greater may be sufficient. Furthermore, it can be shown that

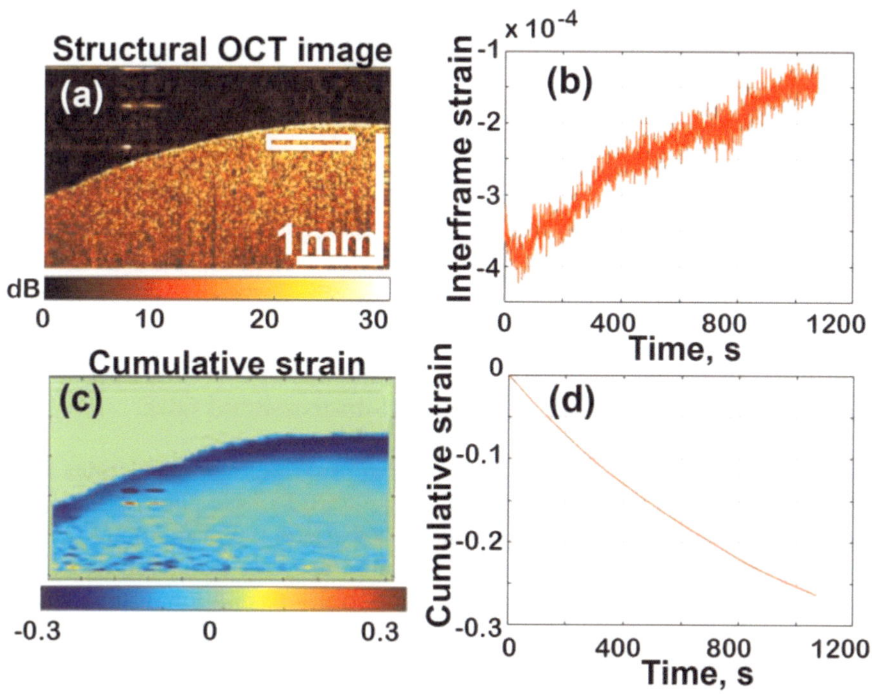

Figure 8.
Monitoring of slow-rate strains arising in the near-surface layer of porcine eye cornea due to its drying in open air. Panel (a) is a typical structural OCT scan of the near-surface cornea region; (b) is the time dependence of interframe strains found with averaging over the rectangle area shown in panel (a); (c) is an example of the spatially resolved map of cumulative strain after 20 min of drying; (d) is the time dependence of cumulative strain corresponding to interframe strain shown in (a).

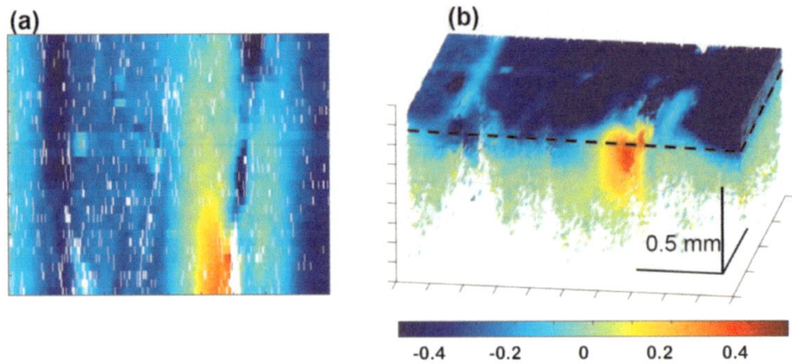

Figure 9.
3D visualization of cumulative strains for a cartilaginous sample (porcine rib cartilage). Drying of the near-surface layer causes its gradual shrinking (negative sign of strain). The presented snapshots correspond a monitoring interval of 145 s. Panel (a) is the horizontal CT scan of cumulative-strain distribution corresponding to the plane marked by dashed lines in an isometric 3D view in panel (b). The localized area with positive strain sign is related to dilatation of the tissue area preliminary subjected to compression by inserting a needle.

from the viewpoint of enhancing SNR in problems of monitoring of slow processes, temporal rarefaction of OCT data may be even more advantageous than acquisition with the maximal rate (see details of the corresponding discussion in [34]).

In such situations, even systems with moderate acquisition rate can be quite sufficient for realization of efficient monitoring of slow deformations. **Figure 8** gives an example monitoring of fairly slow strains for drying cornea of a porcine eye (total duration of the record in 20 min with 1 s time step). The development of pronounced shrinking of the drying near-surface layer is clearly seen.

Figure 9 shows a 3D example of monitoring of slow strain caused by mechanical relaxation of a cartilaginous sample, in which initially a local compression was produced by an inserted needle. In this example the strain was reconstructed via comparison of entire 3D sets of complex-valued OCT data (acquisition of one 3D data set required 1.6 s). The signal processing, as in the other previous examples, was made using the above-described vector method for estimating gradients of interframe phase variations. More detailed discussion of optimizations for the monitoring of slow strains can be found in [34].

5. Obtaining of quantitative stress-strain curves and estimation of Young modulus in phase-sensitive compressional OCE

Consider now possibilities of the developed approach to mapping strains for quantitative mapping of the Young modulus in the studied samples. As was discussed in paper [7] related to ultrasound-based elastography and pointed out in the Introduction to this chapter, if a rigid piston compresses a material that can fairly freely expand in the lateral directions, the stress distribution near the piston can be close to uniaxial. Consequently, the reaction to such compression is determined by the Young modulus of the material. This statement is the basic principle of compression (or compressional) quasistatic elastography [8]. Thus, the Young modulus can be estimated by measuring strain produced by a known stress applied to the sample by a compressing piston. It was also mentioned in paper [7] that a reference pre-calibrated layer overlaying the studied material can be used as a kind of compliant sensor to control the stress by measuring the strain within the reference layer. This idea has not found application in the ultrasound-based compressional elastography (where the compression approach is mostly used for visualization of relative distribution of stiffness); however, the application of a reference layer has proven to be useful in OCT-based quantitative elastography [31, 35]. In [35] special attention was paid to the effects of stiction between the compressing rigid piston (usually, the output window if the OCT probe) and the reference silicone layer. The stiction impedes free lateral expansion of the compressed layer. Consequently, its compressibility becomes lower (in other words, the distorted apparent Young modulus seems to be greater), because under constrained lateral expansion, the layer reaction to loading is determined by some mixture of the Young modulus and much greater bulk modulus (the proportion of their contributions being determined by the strength of stiction). If the possibility to observe inhomogeneity of the local strain inside the translucent reference layer is ensured, this makes it possible to detect the presence of stiction and take the necessary measures to reduce its distorting effect (e.g., to add a lubricant liquid between the compressing solid surface and the silicone layer). For a non-scattering, very transparent reference layer, its strain can be estimated only by observing its total thickness, so that the strain distribution inside the layer is impossible to detect optically. Besides, the accuracy of strain estimation via changes in the total thickness usually is lower, so that the usage of translucent layers is preferable in practice. Panels (a) and (b) in **Figure 10** illustrate the application of a reference layer as a compliant sensor. The interframe phase-variation distribution in **Figure 10b** demonstrates evident inhomogeneity of the vertical phase-variation gradient, although the material of the layer is mechanically homogeneous. The layer region contacting with the stiff surface of the OCT probe looks more stiff because of the silicone-glass stiction. **Figure 10c** and **d** demonstrates the interframe-strain profiles within the mechanically compressed reference silicone layer and the underlying tissue in the experimental configuration shown in **Figure 5** for two different degrees of silicone-glass stiction. In **Figure 10c** pronounced inhomogeneity of strain is visible, which looks

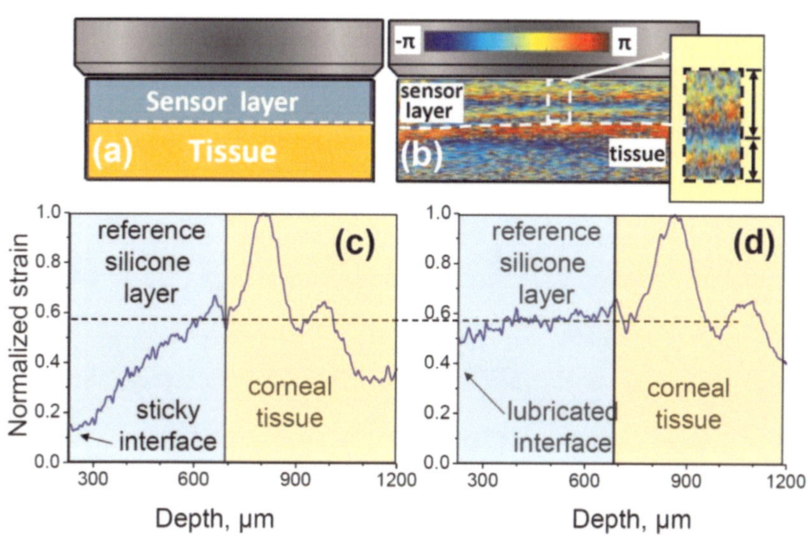

Figure 10.
Elucidation of principle of OCE-based quantitative compressional elastography and potential stiction-related distortions. Panel (a) schematically shows the reference silicone layer over the studied tissue; (b) is a real example of interframe phase variation with inhomogeneous phase-variation gradient within the homogeneous silicone layer; (c) is an example of pronouncedly inhomogeneous phase variation as a function of depth because of noticeable stiction at the silicone-glass interface; (d) is a similar graph obtained in the same configuration after adding a drop of lubricating liquid at the silicone-glass interface for reduction of stiction.

as if the Young modulus of silicone varied ~4 times over the layer thickness. In **Figure 10d** obtained in the same configuration by adding a lubricating liquid between the glass and silicone, the stiction is strongly reduced, so that the distribution of strain within the layer becomes rather homogeneous (as should be expected for the homogeneous silicone). Thus, one should pay due attention to controlling the stiction in order to avoid possible strong distortions in the apparent Young modulus of the reference layer.

If sufficiently low stiction at the interface with the glass is ensured, by measuring strain within pre-calibrated silicone layers, one can estimate stress applied to the studied underlying material. In this context another essential point is whether it is possible to neglect the dependence of the Young modulus of the silicone on the degree of silicone straining. In other words, what is the strain range within which silicone behaves as fairly linear material with stiffness independent of the applied stress? This is an important point, because unlike acoustics with typical strains below $10^{-4} - 10^{3}$, in mechanical tests of polymers and biological samples, their strains may reach several percent and even 10%. In this context, **Figure 11a** demonstrates the results of a kind of "self-calibrating" tests allowing to verify linearity of mechanical behavior of silicones. In this test, sandwich structures composed of silicone layers with different Young moduli were used. The stiffer layer experienced smaller strain for the same stress remaining within an expectedly more linear region. If both compressed materials are linear, then for one strain plotted against another, one should expect a linear function. The curves shown in **Figure 11a** obtained for various combinations of softer/stiffer silicones are fairly linear with a rather broad strain ranging up to several tens of percent. This confirms that silicone is a good candidate for linear reference material in which strain is proportional to stress.

Consequently, if the dependence of strain in a studied tissue plotted against strain in the reference silicone layer looks nonlinear, the reason for this is that the stress-strain dependence for this tissue is nonlinear. For pre-calibrated reference layers, this opens the possibility to obtain quantified stress-strain dependences as illustrated in **Figure 11b**. The example in **Figure 11b** demonstrates the importance of the tissue pre-straining: even for apparently insignificant pre-straining within 2.5%, the apparent Young modulus may vary several times because of possible nonlinearity of the studied sample.

Further, we demonstrate that the described OCT-based technique opens the possibility to observe spatially localized inhomogeneities of the material stiffness.

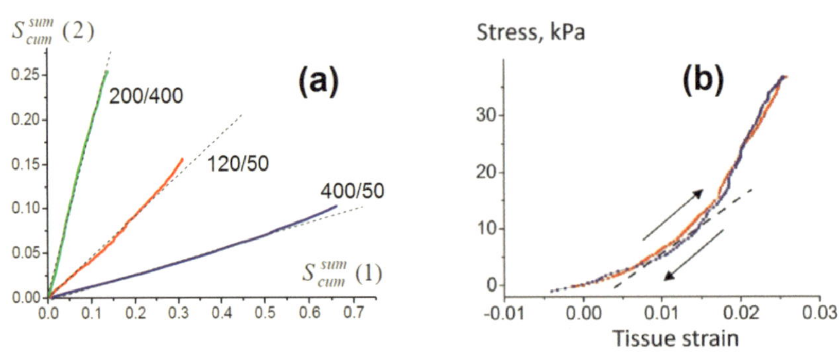

Figure 11.
Panel (a) shows cumulative strain in one silicone layer plotted against strain in another layer. The numbers near the curves indicate the corresponding ratios of the Young moduli for the compared silicone types. The dependences in panel (a) remain fairly linear up to strains of ~10%. Panel (b) shows another example of a pronounced nonlinear stress-strain curve, for which the strain in pre-calibrated silicone is recalculated in stress and the horizontal axis shows the strain in an excised sample of breast cancer. Note that the slope of this curve (corresponding to the Young modulus of the tissue) strongly varies from 400 kPa to 1400 kPa for the strain range in the tissue <2.5%.

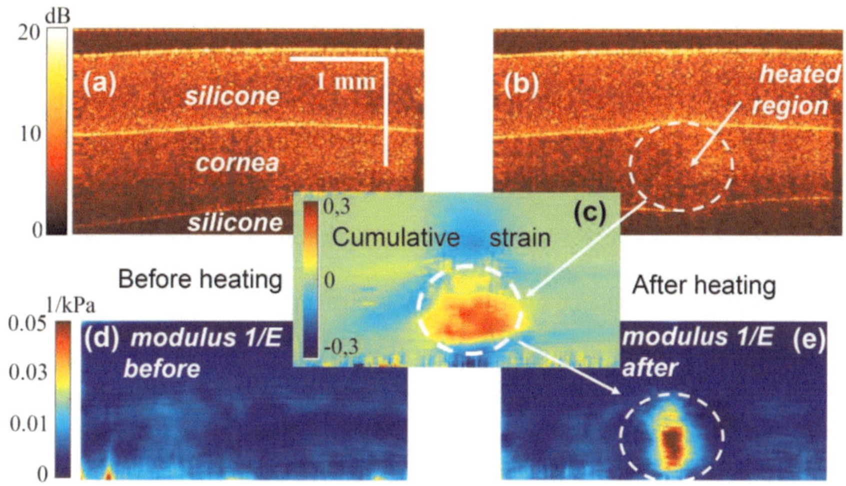

Figure 12.
Demonstration of complementary irradiation-induced residual strain and reduction in the Young modulus in the irradiated region of corneal sample. Panel (a) is a structural OCT image before irradiation, and (b) is a similar postirradiation structural image, in which slight expansion of the corneal tissue is distinguishable in the encircled zone. Panel (c) is an elastographic map of the postirradiation cumulative strain, where expansion of the heated region is clearly visible. Panel (d) is the elastographic map of spatially fairly uniform distribution of the inverse Young modulus (compressibility) before irradiation. Similar map (e) obtained after irradiation clearly demonstrates appearance of local increase in compressibility (reduction in the Young modulus) in the center of irradiation region.

The examples presented in **Figure 12** are also obtained in the experimental configuration shown in **Figure 5**. **Figure 12a** and **b** shows structural images of the silicone-cornea-silicone sandwich before and after pulse-periodic infrared irradiation. The latter causes deformation (expansion) of the irradiated region such that after postirradiation cooling the residual expansion persists. It is slightly visible in the structural image **Figure 12b** and can be quantitatively estimated and clearly visualized as shown in **Figure 12c**. For the pre-compression used in the discussed experiment, the Young modulus of the corneal tissue is close to the Young modulus of the silicone (~200 kPa). Consequently, before the irradiation the reconstructed distribution of the compressibility (inverse Young modulus) over the image found via comparison of spatial distribution of mechanically produced small straining ~0.5% is fairly uniform and does not vary laterally in the cornea (see **Figure 12d**). In contrast, after laser heating, the compressibility of cornea demonstrates strong increase in the center of the heated region (see **Figure 12e**). This local increase in the compressibility of the heated region is attributed to the laser-induced appearance of crack-like microscopic pores that are not directly resolved in the OCT images, but due to enhanced deformability, they manifest themselves via reduction of the tissue Young modulus. Analysis of the post-heating local dilatation of the tissue and the complementary reduction in the Young modulus makes it possible to make quantitative conclusions about the volume content and averaged geometrical parameters of the pores. These results well agree with independent data of microscopic examination of the tissue (see details in [30]).

6. Conclusions

The described elastographic OCT-based approach can be viewed as an optical counterpart of the elastographic approach [7] proposed in medical ultrasound.

However, an important distinction is that in contrast to the initially proposed correlational tracking in ultrasound, in OCT the speckle tracking based on analysis of the scattered signal phase appeared to be more advantageous. The difficulties in realization of correlational speckle tracking in OCT arise because of strain-induced decorrelation of speckle patterns in OCT scans, so that super-broadband sources are required to reduce the strain-induced speckle blinking/boiling as discussed in detail in [19].

It can be shown that the phase-sensitive OCT-based speckle tracking is more tolerant to strains [21], so that phase-resolved tracking of axial displacements can be efficiently realized even using OCT systems with "typical" parameters. In problems of mapping strains, the difficulties in phase unwrapping for super-wavelength displacements of scatterers can be efficiently obviated by direct estimation of interframe phase-variation gradients within a chosen processing window. This can be done using the proposed robust "vector" method [24, 25]. The resolution of the resultant strain maps is mostly determined by the dimensions of the used processing window, the size of which should usually be at least 5–10 times greater (depending on the noise level) than the resolution scale in the initial structural OCT images to ensure sufficient SNR.

Additional ways of SNR enhancement can be based on application of stable periodic actuators producing strain and periodic averaging (e.g., see [11]; however, this solution cannot be used to determine the quasistatic Young modulus). Alternatively in the case of aperiodic strains, efficient averaging can be obtained by finding cumulative strains as discussed in [32], which is also a key point allowing for obtaining nonlinear stress-strain curves as demonstrated above. Additional improvements in terms of optimization of interframe intervals may be required in the case of long-term monitoring of slowly varying small strains [34]. In the context of biomedical applications, some examples related to biomechanics of such biopolymers such as cartilaginous and corneal tissues were given above (a more detailed discussion can be found in [28–30]). Other examples related to oncological applications can be found in [11, 12]. It is likely that monitoring of slow strains (as discussed in [34]) may be of special interest for technical applications related to studies of curing of polymers, etc.

Acknowledgements

The authors acknowledge the support of the Russian Foundation for Basic Research (Grant No. 18-42-520018 in part of the development of 3D data acquisition and Grant No. 18-32-20056 in part of the development of OCE-based reconstruction of nonlinear stress-strain dependences). Obtaining of the experimental data for cartilaginous and corneal samples was supported by the Russian Science Foundation (Grant No. 16-15-10274).

Author details

Vladimir Y. Zaitsev*, Lev A. Matveev, Alexander A. Sovetsky
and Alexander L. Matveyev
Institute of Applied Physics of the Russian Academy of Sciences, Nizhny Novgorod,
Russia

*Address all correspondence to: vyuzai@mail.ru

References

[1] Hill CR, Bamber JC, ter Haar GR, editors. Ultrasound in medicine. In: Physical Principles of Medical Ultrasonics. 2nd ed. J. Surrey, UK: Wiley and Sons; 2004. p. 528. DOI: 10.1002/0470093978

[2] De Carlo TE, Romano A, Waheed NK, Duker JS. A review of optical coherence tomography angiography (OCTA). International Journal of Retina and Vitreous. 2015;**1**(5):1-15. DOI: 10.1186/s40942-015-0005-8

[3] Maslennikova AV, Sirotkina MA, Moiseev AA, Finagina ES, Ksenofontov SY, Gelikonov GV, et al. In-vivo longitudinal imaging of microvascular changes in irradiated oral mucosa of radiotherapy cancer patients using optical coherence tomography. Scientific Reports. 2017;**7**(1):15505 (1-10). DOI: 10.1038/s41598-017-16823-2

[4] Sirotkina MA, Matveev LA, Shirmanova MV, Zaitsev VY, Buyanova NL, Elagin VV, et al. Photodynamic therapy monitoring with optical coherence angiography. Scientific Reports. 2017;7:41506(1-11). DOI: 10.1038/srep41506

[5] Sirotkina MA, Moiseev AA, Matveev LA, Zaitsev VY, Elagin VV, Kuznetsov SS, et al. Accurate early prediction of tumour response to PDT using optical coherence angiography. Scientific Reports. 2019;**9**(1):6492(1-9). DOI: 10.1038/s41598-019-43084-y

[6] Schmitt J. OCT elastography: Imaging microscopic deformation and strain of tissue. Optics Express. 1998;**3**(6):199-211. DOI: 10.1364/OE.3.000199

[7] Ophir J, Cespedes I, Ponnekanti H, Yazdi Y, Li X. Elastography: A quantitative method for imaging the elasticity of biological tissues. Ultrasonic Imaging. 1991;**13**:111-134. DOI: 10.1016/0161-7346(91)90079-W

[8] Parker KJ, Doyley MM, Rubens DJ. Imaging the elastic properties of tissue: The 20 year perspective. Physics in Medicine and Biology. 2011;**56**(1):R1-R29. DOI: 10.1088/0031-9155/57/16/5359

[9] Kennedy BF, Wijesinghe P, Sampson DD. The emergence of optical elastography in biomedicine. Nature Photonics. 2017;**11**(4):215-221. DOI: 10.1038/nphoton.2017.6

[10] Kirby MA, Pelivanov I, Song S, Ambrozinski Ł, Yoon SJ, Gao L, et al. Optical coherence elastography in ophthalmology. Journal of Biomedical Optics. 2017;**22**(12):121720(1-28). DOI: 10.1117/1.JBO.22.12.121720

[11] Allen WM, Kennedy KM, Fang Q, Chin L, Curatolo A, Watts L, et al. Wide-field quantitative micro-elastography of human breast tissue. Biomedical Optics Express. 2018;**9**(3): 1082-1096. DOI: 10.1364/BOE.9.001082

[12] Gubarkova EV, Sovetsky AA, Zaitsev VY, Matveyev AL, Vorontsov DA, Sirotkina MA, et al. OCT-elastography-based optical biopsy for breast cancer delineation and express assessment of morphological/molecular subtypes. Biomedical Optics Express. 2019;**10**(5):2244-2263. DOI: 10.1364/BOE.10.002244

[13] Dong B, Pan B, Zhang Y, Bai Y. Microdefect identification in polymers by mapping depth-resolved phase-difference distributions using optical coherence tomography. Polymer Testing. 2018;**68**(April):233-237. DOI: 10.1016/j.polymertesting.2018.04.023

[14] Bercoff J, Tanter M, Fink M. Supersonic shear imaging: A new technique for soft tissue elasticity mapping. IEEE Transactions on Ultrasonics, Ferroelectrics, and Frequency Control. 2004;**51**(4):

396-409. DOI: 10.1109/TUFFC.2004.1295425

[15] Singh M, Wu C, Liu CH, Li J, Schill A, Nair A, et al. Phase-sensitive optical coherence elastography at 1.5 million A-lines per second. Optics Letters. 2015; **40**(11):2588-2591. DOI: 10.1364/OL.40.002588

[16] Pan B, Qian K, Xie H, Asundi A. Two-dimensional digital image correlation for in-plane displacement and strain measurement: A review. Measurement Science and Technology. 2009;**20**(6):062001-1-062001-06200117. DOI: 10.1088/0957-0233/20/6/062001

[17] Rogowska J, Patel NA, Fujimoto JG, Brezinski ME. Optical coherence tomographic elastography technique for measuring deformation and strain of atherosclerotic tissues. Heart. 2004; **90**(5):556-562. DOI: 10.1136/hrt.2003.016956

[18] Zaitsev VY, Gelikonov VM, Matveev LA, Gelikonov GV, Matveyev AL, Shilyagin PA, et al. Recent trends in multimodal optical coherence tomography. I. Polarization-sensitive OCT and conventional approaches to OCT elastography. Radiophysics and Quantum Electronics. 2014;**57**(1):52-66. DOI: 10.1007/s11141-014-9493-x

[19] Zaitsev VY, Matveyev AL, Matveev LA, Gelikonov GV, Gelikonov VM, Vitkin A. Deformation-induced speckle-pattern evolution and feasibility of correlational speckle tracking in optical coherence elastography. Journal of Biomedical Optics. 2015;**20**(7):075006 (1-12). DOI: 10.1117/1.JBO.20.7.075006

[20] Nahas A, Bauer M, Roux S, Boccara AC. 3D static elastography at the micrometer scale using full field OCT. Biomedical Optics Express. 2013;**4**(10): 2138-2149. DOI: 10.1364/BOE.4.002138

[21] Zaitsev VY, Vitkin IA, Matveev LA, Gelikonov VM, Matveyev AL, Gelikonov GV. Recent trends in multimodal optical coherence tomography. II. The correlation-stability approach in OCT Elastography and methods for visualization of microcirculation. Radiophysics and Quantum Electronics. 2014;**57**(3): 210-225. DOI: 10.1007/s11141-014-9505-x

[22] Müller HH, Ptaszynski L, Schlott K, Debbeler C, Bever M, Koinzer S, et al. Imaging thermal expansion and retinal tissue changes during photocoagulation by high speed OCT. Biomedical Optics Express. 2012;**3**(5):1025-1046. DOI: 10.1364/BOE.3.001025

[23] Kennedy BF, Koh SH, McLaughlin RA, Kennedy KM, Munro PRT, Sampson DD. Strain estimation in phase-sensitive optical coherence elastography. Biomedical Optics Express. 2012;**3**(8):1865-1879. DOI: 10.1364/BOE.3.001865

[24] Zaitsev VY, Matveyev AL, Matveev LA, Gelikonov GV, Sovetsky AA, Vitkin A. Optimized phase gradient measurements and phase-amplitude interplay in optical coherence elastography. Journal of Biomedical Optics. 2016;**21**(11):116005(1-12). DOI: 10.1117/1.JBO.21.11.116005

[25] Matveyev AL, Matveev LA, Sovetsky AA, Gelikonov GV, Moiseev AA, Zaitsev VY. Vector method for strain estimation in phase-sensitive optical coherence elastography. Laser Physics Letters. 2018;**15**(6):065603(1-6). DOI: 10.1088/1612-202X/aab5e9

[26] Zaitsev VY, Matveyev AL, Matveev LA, Gelikonov GV, Gubarkova EV, Gladkova ND, et al. Hybrid method of strain estimation in optical coherence elastography using combined sub-wavelength phase measurements and supra-pixel displacement tracking. Journal of Biophotonics. 2016;**9**(5): 499-509. DOI: 10.1002/jbio.201500203

[27] Zaitsev VY, Matveev LA, Matveyev AL, Gelikonov GV, Gelikonov VM. A model for simulating speckle-pattern evolution based on close to reality procedures used in spectral-domain OCT. Laser Physics Letters. 2014; **11**(10):105601(1-8). DOI: 10.1088/1612-2011/11/10/105601

[28] Zaitsev VY, Matveyev AL, Matveev LA, Gelikonov GV, Omelchenko AI, Shabanov DV, et al. Optical coherence tomography for visualizing transient strains and measuring large deformations in laser-induced tissue reshaping. Laser Physics Letters. 2016; **13**(11):115603(1-8). DOI: 10.1088/1612-2011/13/11/115603

[29] Zaitsev VY, Matveyev AL, Matveev LA, Gelikonov GV, Omelchenko AI, Baum OI, et al. Optical coherence elastography for strain dynamics measurements in laser correction of cornea shape. Journal of Biophotonics. 2017;**10**(11):1450-1463. DOI: 10.1002/jbio.201600291

[30] Zaitsev VY, Matveyev AL, Matveev LA, Gelikonov GV, Baum OI, Omelchenko AI, et al. Revealing structural modifications in thermomechanical reshaping of collagenous tissues using optical coherence elastography. Journal of Biophotonics. 2019;**12**(3):e201800250 (1-15). DOI: 10.1002/jbio.201800250

[31] Kennedy KM, Chin L, McLaughlin RA, Latham B, Saunders CM, Sampson DD, et al. Quantitative micro-elastography: Imaging of tissue elasticity using compression optical coherence elastography. Scientific Reports. 2015;**5**: 15538-1-15538-1553812. DOI: 10.1038/srep15538

[32] Sovetsky AA, Matveyev AL, Matveev LA, Shabanov DV, Zaitsev VY. Manually-operated compressional optical coherence elastography with effective aperiodic averaging: Demonstrations for corneal and cartilaginous tissues. Laser Physics Letters. 2018;**15**:085602(1-8). DOI: 10.1088/1612-202X/aac879

[33] Lawman S, Madden PW, Romano V, Dong Y, Mason S, Williams BM, et al. Deformation velocity imaging using optical coherence tomography and its applications to the cornea. Biomedical Optics Express. 2017;**8**:5579-5593. DOI: 10.1364/BOE.8.005579

[34] Zaitsev VY, Matveev LA, Matveyev AL, Sovetsky AA, Shabanov DV, Ksenofontov SY, et al. Optimization of phase-resolved optical coherence elastography for highly-sensitive monitoring of slow-rate strains. Laser Physics Letters. 2019;**16**(6):065601(1-6). DOI: 10.1088/1612-202X/ab183c

[35] Zaitsev VY, Matveyev AL, Matveev LA, Gubarkova EV, Sovetsky AA, Sirotkina MA, et al. Practical obstacles and their mitigation strategies in compressional optical coherence elastography of biological tissues. Journal of Innovative Optical Health Sciences. 2017;**10**(6):1742006(1-13). DOI: 10.1142/S1793545817420068

Chapter 10

OCT with a Visible Broadband Light Source Applied to High-Resolution Nondestructive Inspection for Semiconductor Optical Devices

Nobuhiko Ozaki, Kazumasa Ishida, Tsuyoshi Nishi, Hirotaka Ohsato, Eiichiro Watanabe, Naoki Ikeda and Yoshimasa Sugimoto

Abstract

Optical coherence tomography with a visible broadband light source (vis-OCT) was developed for high-resolution and nondestructive measurements of semiconductor optical devices. Although a near-infrared (NIR) light source should be used for medical OCT to obtain deep penetration of biological samples, a visible broadband light source is available as a low-coherence light source for industrial products. Vis-OCT provides higher axial resolution than NIR-OCT, because the axial resolution of an OCT image is proportional to the square of the center wavelength of the light source. We developed vis-OCT with an axial resolution of less than 1 μm in air and obtained cross-sectional profiles and images of ridge-type waveguides having heights and widths of several μm. Additionally, we performed cross-sectional measurements and imaging of a stacked semiconductor thin layer. The measured values were similar to those measured by scanning electron microscopy, and the effectiveness of vis-OCT for nondestructive inspection of semiconductor optical devices was demonstrated.

Keywords: vis-OCT, nondestructive inspection, optical device fabrication, semiconductor optical device, high-resolution imaging

1. Introduction

Nondestructive inspection technologies have become mandatory for fabrication and evaluation of various industrial products [1]. In particular, as products are miniaturized into μm and nm scales, high-resolution measurement methods are necessary. For semiconductor optical devices, photonic integrated circuits (PICs) [2] have been developed from the recent progress of fabrication techniques. For the PICs, highly dense optical devices are connected using optical waveguides [e.g., ridge-type optical waveguides (RWGs)] having several to sub-μm scales to propagate light [3, 4]. Thus, the structural accuracy of the fabricated waveguides becomes significant for reducing their propagation loss and increasing PIC efficiency and reliability. The waveguides are typically carved on semiconductor materials using fine-processing technologies such as photolithography and dry etching [5]. To optimize the process conditions, it is necessary to measure the thickness of a photoresist coating on semiconductor wafers: several to sub-μm. Furthermore, the fabricated waveguides should be inspected as designed. To measure the photoresist thickness or fabricated waveguides, scanning electron microscopy (SEM) or step profiler has been frequently used [6]. However, these measurement methods are destructive or require contact with the sample. Development of a nondestructive and non-contact measurement method with a high spatial resolution is thus required.

Optical techniques enable nondestructive, contactless characterization of thin films. For instance, ellipsometry [7] is commonly used to measure thicknesses of thin films. However, it has a relatively large spot size (>40 μm) and is not suitable for a lateral local measurement of wafers. On the other hand, an optical cross-sectional imaging technique, called optical coherence tomography (OCT) [8], has been developed. OCT is commonly used in medical fields (e.g., ophthalmology). It is based on low-coherence interferometry, and the reflectivity profile of a sample along an optical axis can be obtained using a focused probe. By scanning the probe in the lateral directions, two-dimensional (2D) and three-dimensional (3D) profile imaging can be achieved. The axial resolution of an OCT image is governed by the central wavelength (λ_0) and bandwidth ($\Delta\lambda$) of the low-coherence light source. The axial resolution can be expressed as $0.44\lambda_0^2/\Delta\lambda$ when a broadband light source has a Gaussian spectral shape [9]. For conventional medical OCT, λ_0 should be in the near-infrared (NIR) range to allow deep penetration of biological samples [10, 11]. However, when industrial products are observed by OCT, a visible broadband light can be used as an OCT light source, and the axial and lateral resolutions of an OCT image can be further improved beyond that of the NIR-OCT images, owing to the shorter λ_0. For instance, when the λ_0 is 650 nm and $\Delta\lambda$ is beyond 200 nm, the axial resolution can be expected to be less than 1 μm. This value is comparable to the dimension of optical semiconductor devices included in PICs. Furthermore, the visible-range optical components, such as light source and detector, are widely used and are relatively cost-effective, and there are many options. Therefore, we developed vis-OCT as a nondestructive high-resolution inspection tool [12, 13]. In this chapter, the vis-OCT system and its inspections of optical components, semiconductor-based optical waveguides, polymer-based integrated patterns, and semiconductor thin film are introduced.

2. Spectral-domain OCT with a visible broadband light source

A spectral-domain (SD)-OCT [14], having a visible broadband light, was constructed as shown in **Figure 1(a)**. SD-OCT, categorized as a Fourier-domain OCT [15], enables a distribution of spatial reflectivities along the optical axis through an inverse Fourier-transformed (IFT) spectrum of interference among reflections from the sample and a reference mirror. We utilized a halogen lamp unit (ANDO AQ4305) coupled with a single-mode fiber of ~4.0 μm core diameter as the visible broadband light source. The integrated output power from the SMF in the visible spectral range was approximately 220 nW. The introduced light was collimated using an objective lens (f = 11.7 mm) and split by a beam splitter (BS) into sample and reference arms. Each split light was focused on the sample and the reference

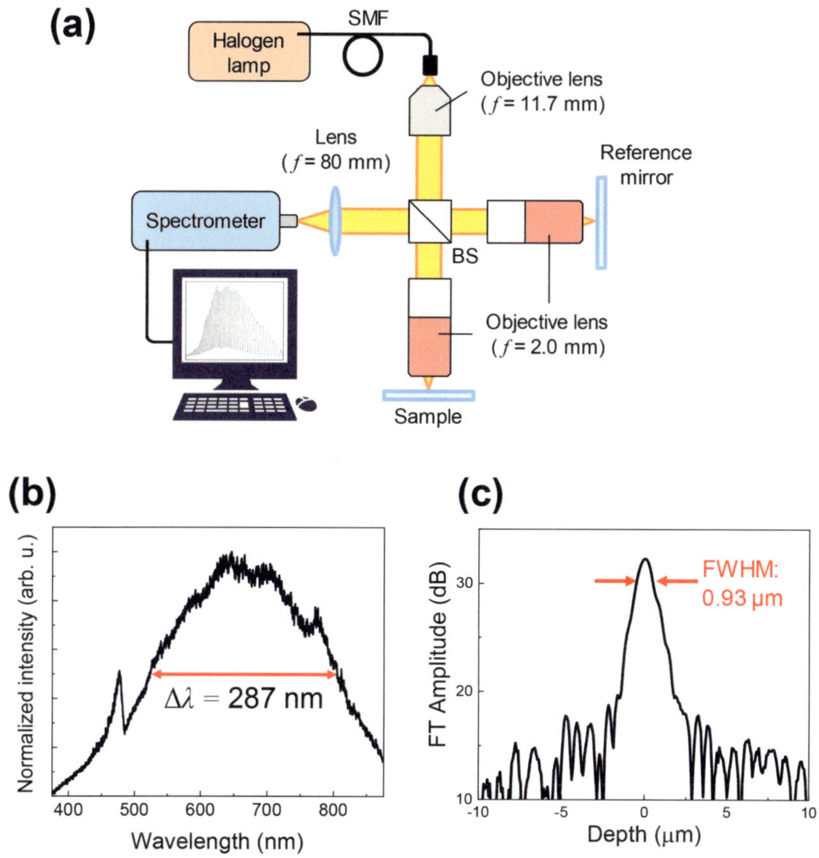

Figure 1.
(a) Schematic of the Vis-OCT setup. (b) Spectrum of the light source in the Vis-OCT. (c) PSF obtained from the IFT interference spectrum between the reflections from the reference and sample mirrors. (Reprinted with permission from Ref. [13]. Copyright 2018, The Japan Society of Applied Physics).

mirror by a high numerical aperture (0.50) objective lens with a plan apochromatic aberration correction from visible to near-infrared wavelength region (Mitutoyo M Plan Apo NIR 100×) (f = 2.0 mm). By using this objective lens, the spot size of each focused beam could be reduced, estimated to be approximately 0.68 μm in diameter without degradation of the axial resolution caused by aberrations. The reflections from the sample and the reference mirror were recombined at the BS, and the interfered signal was focused by an achromatic lens (f = 80 mm) into the spectrometer. The interference spectrum was detected by the spectrometer (Ocean Optics USB4000) with a Si-based charge-coupled device detector (3648 pixels) with an exposure time of approximately 100 ms. By calculating the IFT interference spectrum, a depth profile of the sample along the optical axis was obtained [16]. A 2D OCT image was then produced by scanning the probe light in steps of 0.2 μm in the lateral direction.

Figure 1(b) shows a spectrum of the light source reflecting from the reference mirror detected with the spectrometer. The center wavelength was approximately 662 nm, and the full width at half maximum (FWHM) was approximately 287 nm. The axial resolution of the SD-OCT system can be estimated from the point spread function (PSF) of the light source obtained from the IFT interference spectrum between the reflections from the reference mirror and an identical mirror set on the sample stage. The PSF of the light source is presented in **Figure 1(c)**, and the FWHM of the PSF, which corresponds to the axial resolution, was approximately 0.93 μm.

3. Surface-structure observations and measurements of RWGs and periodic patterns with Vis-OCT

We first measured optical waveguides and periodic patterns as typical optical device components to evaluate the axial and lateral resolutions of developed vis-OCT. In this section, the performance of the vis-OCT for surface-structure measurement and observation is described.

3.1 Sample preparations

For the test samples of surface-structure observations, we prepared semiconductor-based straight RWGs. The RWG was formed on typical optical confinement layers, which consisted of a core layer of GaAs and upper- and lower-cladding layers of AlGaAs [17]. After the epitaxial growth of the semiconductor layers on a GaAs substrate via molecular beam epitaxy (MBE), RWGs were separately fabricated on the wafer with 1.5-μm nominal height and various nominal widths (3–100 μm) using conventional photolithography and dry etching processes (**Figure 2**).

We prepared periodic patterned structures fabricated on a soft mold made of polydimethylsiloxane (PDMS) with several to sub-μm scale for nano-imprint lithography (NIL) [18], as shown in **Figure 3**. Three laterally periodic patterns, line and space (L/S), hexagonal lattice pillars, and hexagonal lattice holes, were formed on the soft mold with nominal 4-μm pitch and 0.5-μm height (or depth). These are typically designed for optical micro- and nano-waveguides, photonic crystal structures, or micro-channels for lab-on-a-chip, etc. We investigated the capability of the vis-OCT with these high-dense patterns for lateral and axial high-resolution measurements.

The light probe was introduced from the above sample surface and scanned in-plane direction across the RWG or periodic patterns to obtain the depth profile and profile images.

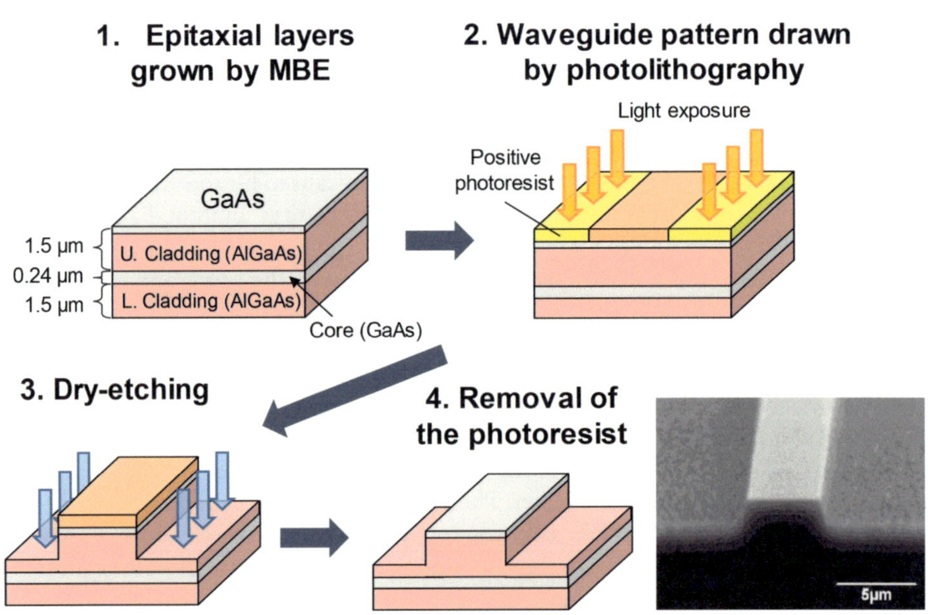

Figure 2.
Schematic of semiconductor RWG fabrication processes and a typical SEM image of the fabricated RWG with a nominal width of 3 μm. (Reprinted with permission from Ref. [13]. Copyright 2018, The Japan Society of Applied Physics).

3.2 Height and width measurements of isolated RWGs

Figure 4(a) shows depth profile curves obtained with a probe fixed on the RWG (black line) and the substrate aside the RWG (red line) [13]. Each profile indicates a peak with a line width of approximately 0.95–0.96 μm, which can be caused by the reflection from the surfaces of the RWG and the substrate, and the line width almost corresponds to that of the PSF of the light source, as shown in **Figure 1(c)**. The peak intensity at the surface of the substrate was smaller (approximately −5 dB) than that at the surface of the RWG. This might have resulted from a reduction in the back reflection caused by light scattering at the dry-etched rough surface. The peak positions at the surfaces were shifted, suggesting that the RWG height can be estimated from the peak-shift value. From this case, the RWG height was estimated at ~1.52 μm. We then obtained the depth profile curves by scanning with the light probe at intervals of 0.2 μm in the lateral direction across the RWG with 3-μm width and summarized the profile curves in **Figure 4(b)**.

As seen in **Figure 4(b)**, the reflection peaks at the surfaces of the substrate and the RWG can be clearly distinguished. At the boundary between the RWG and the substrate, the incident light probe spanned both surfaces, and the signal intensities were overlapped in a short range of the lateral direction. Thus, we defined the boundary position where the peak intensity dropped 3 dB from the peak intensity of the RWG surface. Based on the depth profile curves, signal intensity was scaled

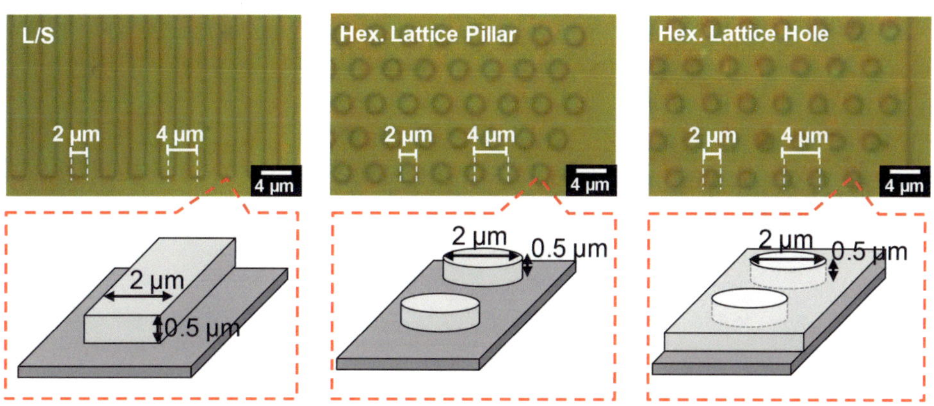

Figure 3.
Optical microscopic and schematics of soft molds for NIL with various lateral patterns.

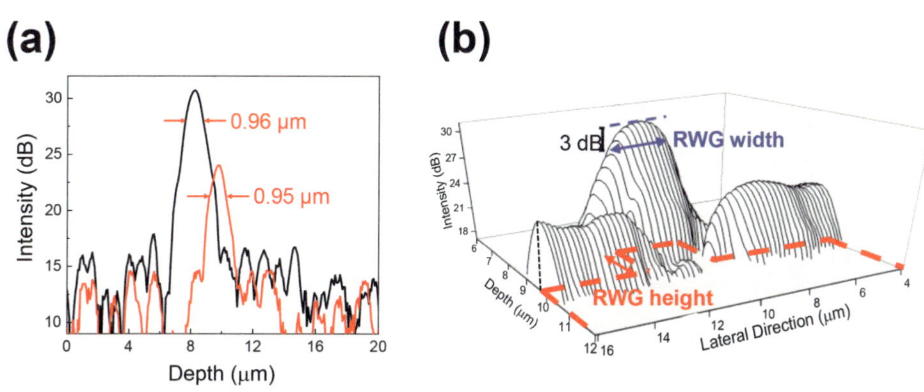

Figure 4.
(a) Depth profile curves obtained from the surface of the RWG with 3-μm width (black line) and the substrate (red line). (b) Summarized profile curves. (Reprinted with permission from Ref. [13]. Copyright 2018, The Japan Society of Applied Physics).

on an 8-bit grayscale, where white denotes maximum and black denotes minimum, and a cross-sectional OCT image was constructed. **Figure 5** summarizes the OCT images obtained from RWGs with various widths and cross-sectional SEM images of the RWGs obtained from a cleaved edge. The surface line of the substrate and the RWG indicated by a red dashed line in each OCT image was determined at the reflection peak position. The RWG height and width measured by vis-OCT correspond well with the values measured via SEM observation. **Figure 6(a), (b)** shows the comparison of the measured heights and widths of various RWGs between vis-OCT and SEM.

3.3 Soft mold for NIL

We then observed soft molds with periodically integrated patterns, such as line and space, hexagonal lattice of pillars, and holes. **Figure 7** shows optical microscopic (plan-view) and vis-OCT (cross-sectional) images of the molds. The OCT images clearly show the profile image of the periodic patterns, indicating the height and pitch of the patterns. Despite the slight variation of dimensions, the measured height and depth of approximately 0.5 μm and the pitch of approximately 4 μm were reasonable values.

These results demonstrated the effectiveness of the vis-OCT for measurements and imaging of integrated structures with several to sub-μm scale. Furthermore, the mold is made of polymer (PDMS) and difficult to be observed by SEM because of charging. Vis-OCT can avoid the problem and realize nondestructive and high-resolution inspection even for the polymer-based material.

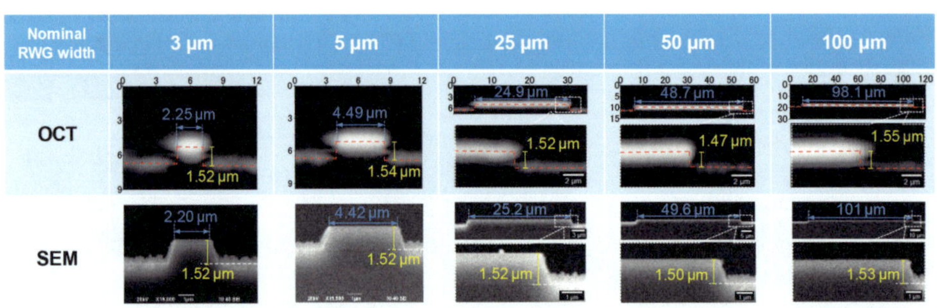

Figure 5.
Comparison of profile images using Vis-OCT and SEM for RWGs with various widths.

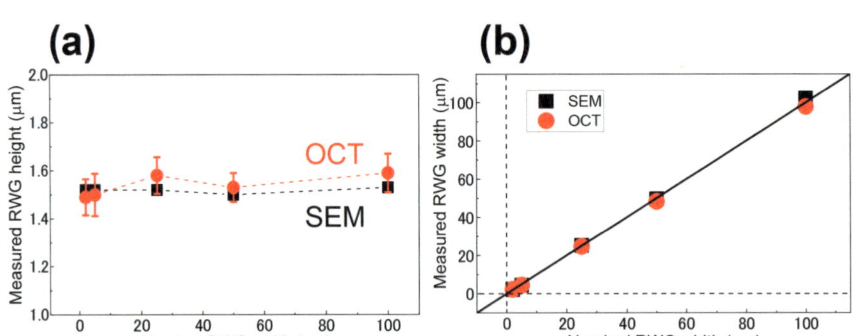

Figure 6.
Summarized measured values of (a) height and (b) width of RWGs with various nominal widths using Vis-OCT and SEM. (Reprinted with permission from Ref. [13]. Copyright 2018, The Japan Society of Applied Physics).

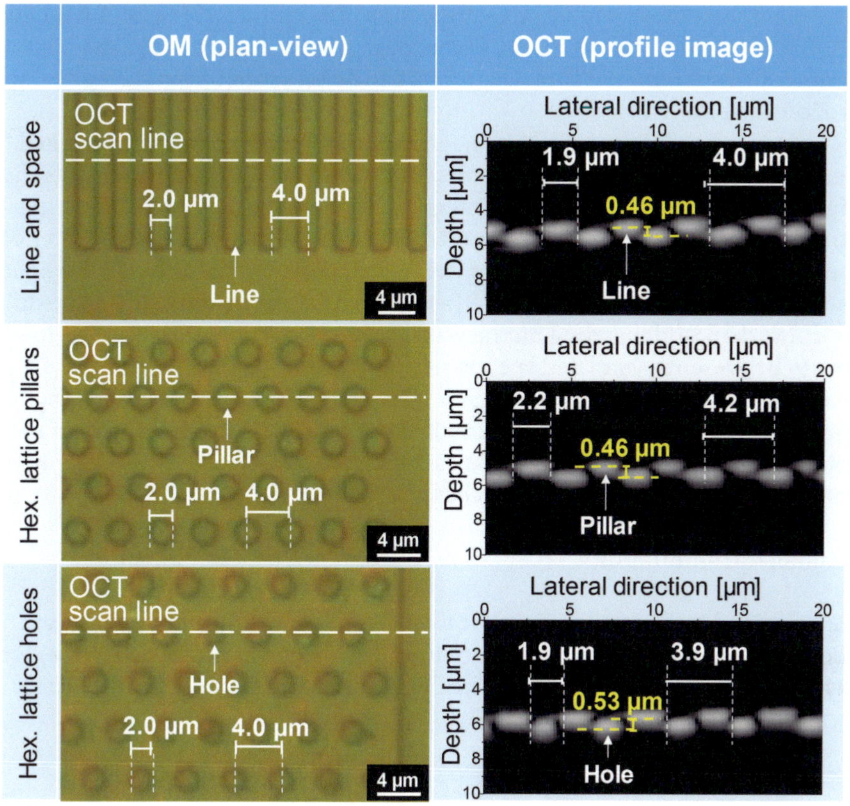

Figure 7.
NIL soft-mold profile images obtained via Vis-OCT and plan-view images with an optical microscope.

4. Profile imaging of semitransparent thin films

As shown in the previous section, the availability of the vis-OCT was demonstrated for nondestructive inspection for surface structures with high axial resolution (less than 1 μm) and lateral resolution (less than 2 μm). We then attempted to obtain a profile imaging of stacked layers underneath the sample surface. Generally, thin semiconductor layers are stacked in optical devices, and an inspection method for the layer embedded in a device, such as the thickness or interface flatness of the layers, is beneficial. Although a layer of transparency for visible light can be easily imaged with vis-OCT, a nontransparent or partial transparent layer, which is an optical absorbent layer for a whole or partial of the visible light, should be difficult to be measured. In this section, our proposed method to determine the physical thickness of such a layer using the vis-OCT is introduced.

4.1 Sample of stacked thin layers

An epitaxial AlGaAs thin layer was grown on a GaAs wafer by MBE as a test sample. As shown in **Figure 8**, a nominal 0.5-μm-thick $Al_{0.35}Ga_{0.65}As$ layer was prepared and was coated with an approximately 2-μm-thick layer of photoresist (AZ 5214E, Microchemicals GmbH). The cross-sectional SEM image was obtained from a cleaved sample edge. A photoresist layer is typically spin-coated before patterns are drawn with a photolithography process. The measurement of thickness of the photoresist layer is important for optimizing the condition of pattern drawing. The refractive index of the photoresist layer (n_{pr}) was approximately 1.62 for

625-nm light, and there is not a large dispersion of n_{pr} for visible light (1.61–1.66 for λ = 830–400 nm) [19], as shown in the lower part of **Figure 9(a)**. The physical thickness of the photoresist film, therefore, can be estimated from the optical thickness divided by n_{pr}.

Measuring the AlGaAs layer thickness is also important, because the etched AlGaAs layer thickness should be monitored when the RWG is fabricated, as shown in **Figure 2**. However, the refractive index of AlGaAs (n_{AlGaAs}) has a large dispersion, unlike the photoresist layer. The refractive index of the $Al_{0.3}Ga_{0.7}As$ layer varies in the range of 3.5–4.8 as a function of the wavelength of visible broadband light, as shown in the upper part of **Figure 9(a)**. Moreover, the $Al_{0.3}Ga_{0.7}As$ layer exhibits optical absorbance, depending on the wavelength, because of the bandgap energy. **Figure 9(b)** presents simulated spectra of the visible broadband light source transmitted through an $Al_{0.3}Ga_{0.7}As$ layer with various thicknesses in the range of 250–2000 nm. This indicates variations of the transmitted spectral shape of the visible broadband light with the thickness of the $Al_{0.3}Ga_{0.7}As$ layer resulting in the variation of n_{AlGaAs} with the spectral changes.

4.2 Numerical simulations

As described in the previous section, the semiconductor layer has a dispersion relation of its refractive index, and the transmitted spectrum is varied with the thickness of the layer. Thus, it should be difficult to determine the physical thickness of the

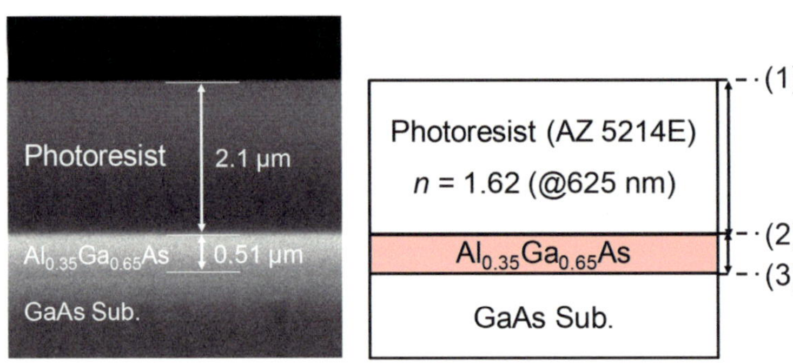

Figure 8.
Cross-sectional SEM image and profiles of a fabricated sample. (Reprinted with permission from Ref. [12]. Copyright 2016, The Japan Society of Applied Physics).

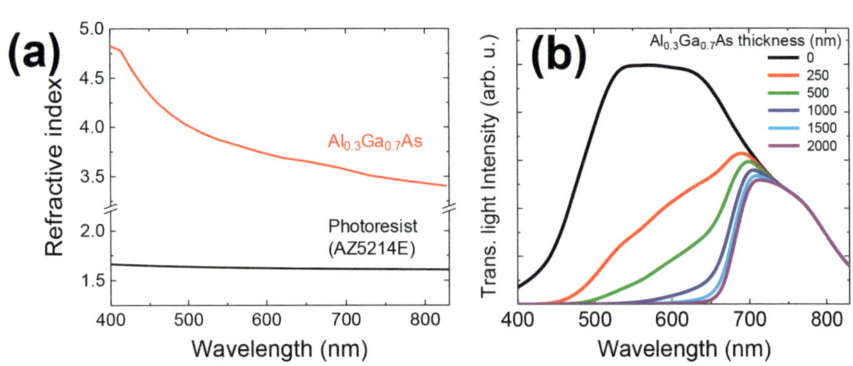

Figure 9.
(a) Dispersion refractive indices of $Al_{0.3}Ga_{0.7}As$ and photoresist. (b) Simulated transmitted spectra through $Al_{0.3}Ga_{0.7}As$ films with various thicknesses. (Reprinted with permission from Ref. [12]. Copyright 2016, The Japan Society of Applied Physics).

AlGaAs layer from a value of optical thickness obtained with vis-OCT. To determine the physical thickness of thin semiconductor layers, we propose the utilization of numerical simulations based on the finite-difference time-domain (FDTD) method [20, 21]. This simulation provides the propagation of an optical wave of visible broadband light in the sample. The FDTD method is a time-domain simulation for modeling electrodynamics and is useful for predictions of light propagation in a wide frequency range in an arbitrary material. The variations in intensity and shape of a spectrum as broadband light travels in the material can be numerically reproduced based on successive calculations of Maxwell's equations in time and space. This enables the estimation of a precise optical length of a broadband light beam traveling in a material having wavelength dispersion of optical absorbance and refractive index.

We executed the FDTD simulation on a 2D model using commercial software (Rsoft Design Group, FullWAVE). As shown in **Figure 10(a)**, a 2D model of stacked layers was prepared, and a pulse light was introduced from the left side of the model. **Figure 10(b)** shows plots of intensity of the incident pulse recorded as a function of time (converted to optical length) at each layer boundary (#1–3). The appearance of peaks in the plot indicates that the light propagated, and the distance between two peaks indicates a traveling length of light corresponding to an optical thickness of the layer. We calculated the optical thickness for simulation models with different physical thicknesses of the AlGaAs layer set around 500 nm. The calculated optical thickness was then compared with the optical thickness experimentally obtained via vis-OCT. The physical thickness was determined as a corresponding thickness set in the simulation model.

4.3 Profile measurements and imaging with the Vis-OCT

Figure 11(a) shows the depth profile of the sample obtained with the vis-OCT. Three clear peaks appear in the profile, which can be attributed to reflections at the boundaries of each layer: (1) the surface of the photoresist, (2) the interface between the photoresist and AlGaAs layer, and (3) the interface between the AlGaAs layer and the GaAs substrate. The distance between peaks (1) and (2) of the sample is 3.69 μm, which is nearly the same as the product (3.4 μm) of the photoresist thickness (2.1 μm) and its approximate n_{pr} of 1.62. Considering that the OCT depth profiles indicate the optical path lengths, peaks (1) and (2) can be interpreted as reflections at the surface and at the interface of the photoresist layer. However, the distance between peaks (2) and (3) cannot be analyzed in the same manner, because the refractive index of AlGaAs is not predictable, as described in the previous section. We thus compared the experimental value with the FDTD simulations for models having various values of physical thickness of the AlGaAs layer. The simulation result, which provided the most comparable optical length, is shown in **Figure 11(b)**. The physical thickness of the AlGaAs layer was set at 512 nm in

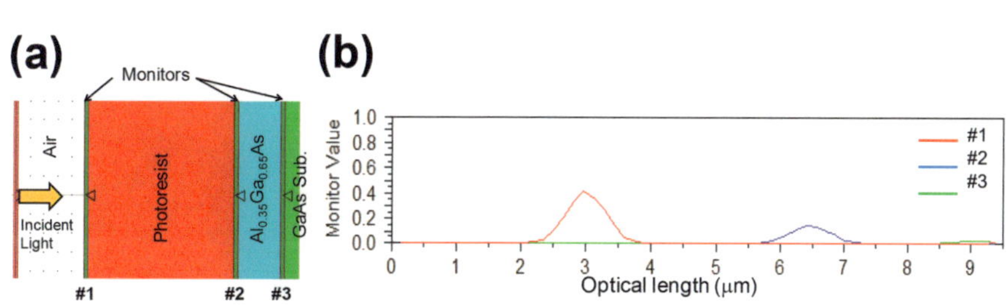

Figure 10.
(a) 2D model for the FDTD simulation. (b) Example of the recorded incident light intensity at layer boundaries #1–3. (Reprinted with permission from Ref. [12]. Copyright 2016, The Japan Society of Applied Physics).

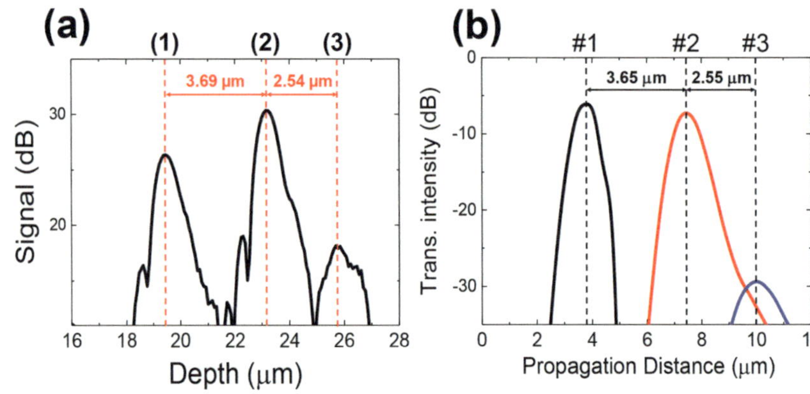

Figure 11.
(a) Depth profile of the sample obtained via Vis-OCT. (b) FDTD simulation result obtained from the most comparable optical length to the experimentally measured profile. (Reprinted with permission from Ref. [12]. Copyright 2016, The Japan Society of Applied Physics).

the simulation model, and the propagation distance (optical length) between layer boundaries #2 and #3 was 2.55 μm. The physical thickness of 512 nm in the model is close to the value measured by the cross-sectional SEM (0.51 μm), as shown in **Figure 8**, demonstrating the effectiveness of the proposed method.

The above results show that vis-OCT, combined with the FDTD simulation, is useful for the profile measurement of not only a transparent film but also the optically absorbent epitaxial semiconductor layer. Although the sample cleaved for the SEM observation was the same wafer used in the vis-OCT measurements, the site for vis-OCT measurement was not at exactly the same position. Thus, the thickness of the grown layer might have been slightly different. Additionally, the visible broadband light spectrum used in the FDTD simulation had an ideal Gaussian shape (symmetric at the central wavelength), which is different from the actual spectrum of the light source used in vis-OCT. This difference in spectral shape should have affected the simulation result, and further optimization of the FDTD simulation parameter could have increased the accuracy.

Furthermore, we obtained a 2D profile image of the sample by collecting profiles in a lateral direction, as shown in **Figure 12**. This image clearly shows the cross-sectional structure of the sample, which could be utilized for inspecting the homogeneity of the film thickness and lateral structures after the etching process. In particular, the clear visualization of the boundary between the AlGaAs and GaAs layers should be useful when a ridge-type waveguide is fabricated on the wafer, because precise measurements of the height of the waveguide and the distance between the bottom

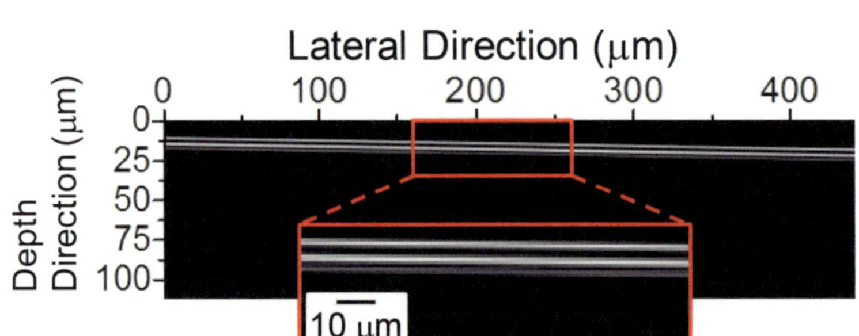

Figure 12.
Profile image of the sample with the Vis-OCT. (Reprinted with permission from Ref. [12]. Copyright 2016, The Japan Society of Applied Physics).

of the waveguide and the GaAs active layer are important requirements for attaining high transmission efficiency in the waveguide. These results demonstrate the effectiveness of vis-OCT for high-resolution, nondestructive profile measurement and imaging in the fabrication of semiconductor optical devices.

5. Conclusions

We demonstrated a nondestructive and non-contact inspection for semiconductor optical devices using vis-OCT with high axial and lateral resolutions of less than 1 μm. Profiles of RWGs and patterned soft molds were obtained, and measurements of the heights and widths of the structures were performed. Furthermore, the physical thickness measurements of an opaque epitaxially grown AlGaAs layer were demonstrated by combining the vis-OCT with FDTD-based numerical simulation. These results offer a novel application of OCT for nonmedical fields.

Acknowledgements

This work was partly supported by the Matching Planner Program sponsored by the Japan Science and Technology Agency (JST). The sample fabrications were supported by the NIMS Nanofabrication Platform in the "Nanotechnology Platform Project" sponsored by the Ministry of Education, Culture, Sports, Science and Technology (MEXT), Japan.

Conflict of interest

The authors declare no conflict of interest.

Author details

Nobuhiko Ozaki[1]*, Kazumasa Ishida[1], Tsuyoshi Nishi[1], Hirotaka Ohsato[2], Eiichiro Watanabe[2], Naoki Ikeda[2] and Yoshimasa Sugimoto[2]

1 Wakayama University, Wakayama, Japan

2 National Institute for Materials Science, Tsukuba, Ibaraki, Japan

*Address all correspondence to: ozaki@wakayama-u.ac.jp

References

[1] Dwivedi SK, Vishwakarma M, Soni A. Advances and researches on non destructive testing: A review. Materials Today: Proceedings. 2018;**5**:3690-3698. DOI: 10.1016/j.matpr.2017.11.620

[2] Coldren LA, Corzine SW, Mashanovitch ML. Diode Lasers and Photonic Integrated Circuits. 2nd ed. Hoboken, NJ: Wiley; 2012

[3] Rickman A. The commercialization of silicon photonics. Nature Photonics. 2014;**8**:579-582. DOI: 10.1038/nphoton.2014.175

[4] Pogossian SP, Vescan L, Vonsovici A. The single-mode condition for semiconductor rib waveguides with large cross section. Journal of Lightwave Technology. 1998;**16**:1851-1853. DOI: 10.1109/50.721072

[5] Xiao H. Introduction to Semiconductor Manufacturing Technology. 2nd ed. Bellingham, WA: SPIE Press; 2012

[6] Stout KJ, Blunt L. Three-Dimensional Surface Topography. 2nd ed. London: Penton Press; 2000

[7] Tompkins HG, Irene EA. Handbook of Ellipsometry. Norwich, NY: William Andrew publishing; 2005

[8] Huang D, Swanson EA, Lin CP, Schuman JS, Stinson WG, Chang W, et al. Optical coherence tomography. Science. 1991;**254**:1178-1181. DOI: 10.1126/science.1957169

[9] Brezinski ME. Optical Coherence Tomography: Principles and Applications. New York: Academic Press; 2006

[10] Patterson MS, Wilson BC, Wyman DR. The propagation of optical radiation in tissue. II: Optical properties of tissues and resulting fluence distributions. Lasers in Medical Science. 1991;**6**:379-390. DOI: 10.1007/BF02042460

[11] Salomatina E, Yaroslavsky AN. Evaluation of the in vivo and ex vivo optical properties in a mouse ear model. Physics in Medicine and Biology. 2008;**53**:2797-2807. DOI: 10.1088/0031-9155/53/11/003

[12] Nishi T, Ozaki N, Oikawa Y, Miyaji K, Ohsato H, Watanabe E, et al. High-resolution and nondestructive profile measurement by spectral-domain optical coherence tomography with a visible broadband light source for optical-device fabrication. Japanese Journal of Applied Physics. 2016;**55**:08RE05. DOI: 10.7567/JJAP.55.08RE05

[13] Ishida K, Ozaki N, Ohsato H, Watanabe E, Ikeda N, Sugimoto Y. Non-destructive and non-contact measurement of semiconductor optical waveguide using optical coherence tomography with a visible broadband light source. Japanese Journal of Applied Physics. 2018;**57**:08PE03. DOI: 10.7567/JJAP.57.08PE03

[14] De Boer JF. Spectral/Fourier domain optical coherence tomography. In: Drexler W, Fujimoto JG, editors. Optical Coherence Tomography: Technology and Applications. New York: Springer; 2008. pp. 147-175

[15] Fercher AF, Hitzenberger CK, Kamp G, El-Zaiat SY. Measurement of intraocular distances by backscattering spectral interferometry. Optics Communication. 1995;**117**:43-48. DOI: 10.1016/0030-4018(95)00119-S

[16] Shibata H, Ozaki N, Yasuda T, Ohkouchi S, Ikeda N, Ohsato H, et al. Imaging of spectral-domain optical coherence tomography using a superluminescent diode based on InAs

quantum dots emitting broadband spectrum with Gaussian-like shape. Japanese Journal of Applied Physics. 2015;**54**:04DG07. DOI: 10.7567/JJAP.54.04DG07

[17] Yasuda T, Ozaki N, Shibata H, Ohkouchi S, Ikeda N, Ohsato H, et al. Electrically driven near-infrared light source with Gaussian-like broadband spectral shape by using multiple self-assembled InAs quantum dots with controlled emission wavelengths. IEICE Transactions on Electronics. 2016;**E99-C**:381-384. DOI: 10.1587/transele.E99.C.381

[18] Chou SY, Krauss PR, Renstrom PJ. Imprint lithography with 25-nanometer resolution. Science. 1996;**272**:85-87. DOI: 10.1126/science.272.5258.85

[19] MicroChemicals photoresists product spec. Available from: http://www.microchemicals.com/technical_information/photoresists_optical_parameters.pdf

[20] Yee K. Numerical solution of initial boundary value problems involving Maxwell's equations in isotropic media. IEEE Transactions on Antennas and Propagation. 1966;**14**:302-307. DOI: 10.1109/TAP.1966.1138693

[21] Taflove A. Application of the finite-difference time-domain method to sinusoidal steady-state electromagnetic-penetration problems. IEEE Transactions on Electromagnetic Compatibility. 1980;**22**:191-202. DOI: 10.1109/TEMC.1980.303879

Chapter 11

Optical Coherence Tomography for Non-Contact Evaluation of Fastener Flushness

James H. Wang and Michael R. Wang

Abstract

Outside of the medical field, spectral domain optical coherence tomography (SD-OCT) is sparsely used. As such, we explored the possibility and practicality of using SD-OCT as a tool to evaluate fastener flushness and countersink surface profiles. A SD-OCT device was built with a handheld galvanometer scanner that weighed only 0.5 lb. Not only it does not require scan center alignment, but it is also capable of quickly producing measurements of fastener flushness, radius, slant angle, countersink edge radius, and surface angle. With the *X-Y* two-line scanning method, measurements take only 90 ms. The SD-OCT device used to obtain these measurements uses a lens with 60 mm focal length and a broadband light source of 840 nm center wavelength and 45 nm spectral bandwidth. With these components, the SD-OCT device is able to provide an axial depth resolution of 8.5 μm and a lateral resolution of 19 μm. The axial depth resolution can be improved by using a wider bandwidth light source. Furthermore, the device is able to produce 3D surface profiles of fasteners and countersinks using multi-line scans.

Keywords: optical coherence tomography, fastener flushness, countersink, optical imaging, 3D surface profile

1. Introduction

High-speed objects such as racecars, bullet trains, aircrafts, and space crafts are dependent on flush surfaces in order to reach the desired aerodynamic performance. As such, inspection techniques are especially important to ensure fastener flushness. Available technology used to inspect fastener flushness and countersink surface profiles include handheld calipers and its modified version [1], laser line scanners [2], and structured light [3]. The contact caliper inspection suffers from gauge contact positioning error, visual gauge angular alignment error, measurement readout error, and inability of producing an instant inspection report. It is also difficult to translate the measurement data into a 3D surface profile. The laser line scanner measurement is a faster technique for accessing the fastener flushness, but it suffers from limited measurement precision and 3D surface profiling ability. The structured light measurement, in particular the fastCHECK introduced by 8tree Inc., offers impressive 5 μm axial measurement resolution and can check the flushness of multiple fasteners. However, it is not capable of 3D surface profiling inspection of fastener tilting angle, fastener radius, and countersink surface angle.

IntechOpen

Optical coherence tomography (OCT) is a low-coherence interferometry technique [4]. With the capability of high axial resolution and fast lateral scan imaging, OCT has been used to perform cross-sectional imaging of biological tissues [5] as well as topographic evaluation of surface roughness [6]. OCT has also been used for fingerprint acquisition [7, 8], inspection of fiber coils [9], surface and coating evaluation [10], evaluation of metallic material fractures [11], and nondestructive metrology of layered polymeric material [12]. We report the use of the spectral domain OCT (SD-OCT) [13–16], a faster OCT, for fastener flushness and countersink surface profile evaluation.

2. SD-OCT configuration, scanning, and resolution

SD-OCT utilizes a Michelson interferometer configuration to examine the interference between two split broadband light beams [4, 14, 15]. **Figure 1** shows the fiber-optic-based SD-OCT configuration which employs a superluminescent diode (SLD) light source; a fiber isolator; a 2 × 2 fiber coupler; a fiber-coupled OCT spectrometer; a reference arm with a collimation lens, a focusing lens, and a reflection mirror; and a measurement arm with a collimation lens, a mirror, galvanometer scanners, and a focusing lens. The SLD light is split by the fiber coupler to illuminate the reference mirror and the fastener as shown in **Figure 1**. The surface scattered light from the fastener and the reflected light from the reference arm mirror are combined by the 2 × 2 fiber coupler to form spectral-dependent interference signals at the OCT spectrometer. The signals are acquired and processed by Fourier transform to extract the fastener surface depth (Z) scattering information, which is the A-scan depth profile information. A lateral scan (B-scan) on X direction on the fastener will yield a fastener surface X-Z profile in the scanned line direction. A multi-line scan (C-scan), which takes multiple X scan lines at different Y positions, can produce a 3D (X-Y-Z) surface profile of the measurement sample. When the fastener is absent in **Figure 1**, the SD-OCT can be used to evaluate the 3D countersink surface profile for countersink production quality inspection.

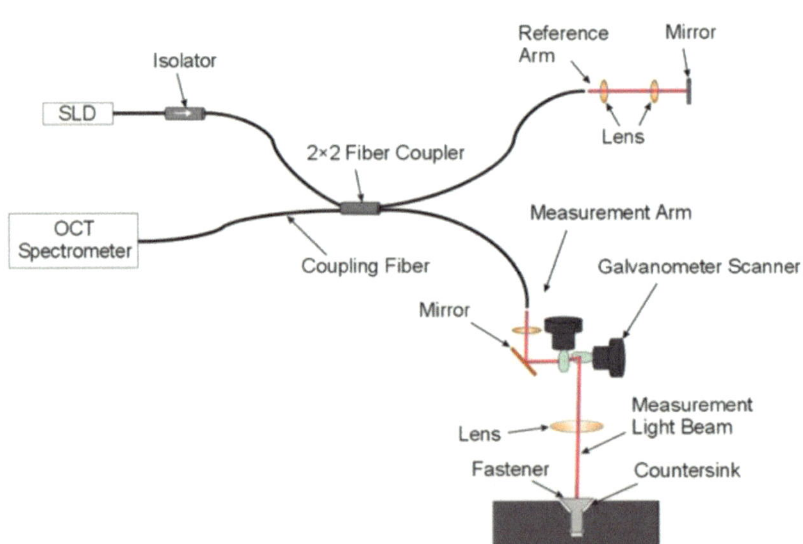

Figure 1.
Schematic of SD-OCT system for fastener flushness evaluation.

We constructed a SD-OCT with a handheld scanner of only 0.5 lb. in weight as shown in **Figure 2** to facilitate easier fastener flushness evaluation. The schematic of the interior of the handheld scanner is also shown. The circular ring will have soft contact (using a mounted plastic or rubber ring) with the surrounding surface to help the user stabilize and align the scanner properly. As to be discussed in the following sections, the scanner circular ring does not need to be centered with the fastener or countersink. An *X* and *Y* two-axis scan would automatically determine the fastener or countersink center and evaluate the fastener slant angle or countersink surface angle.

The SD-OCT system (BIOptoscan OS-184, New Span Opto-Technology) is constructed in-house. An SLD (IPSDD0808, InPhenix) light source is used which has a center wavelength of 840 nm and a full width at half maximum spectral bandwidth of 45 nm. The selection of this light source is due to its availability in-house. For fastener flushness evaluation, it would be desirable in principle to select shorter SLD center wavelength yet with reasonable wide spectral bandwidth to attain a better axial resolution [17]. However, the choice of SLD (e.g., InPhenix) is quite limited. The OCT spectrometer using a line scan camera with 2048 pixels supports an axial depth measurement range of 3.65 mm in the air. The two-axis handheld galvanometer scanner in **Figure 2** is connected to the SD-OCT main unit through a control cable with a single-mode optical fiber. A data acquisition card (PCI-6251, National Instruments) and a camera link card (PCIe-1427, National Instruments) are used. The SD-OCT is operating at a line scan rate of 28 kHz. The Fourier transform data processing is performed by a computer. The handheld scanner's circular ring has a diameter of 3.6 cm, and the focusing lens has a focal length of 60 mm, which produces a focused beam spot size of 19 μm on the sample.

For fast scan measurement, the line scan uses 512 measurement spot positions with 19 μm scan spot center-to-center separation, producing no gap between adjacent spots. The field of view (FOV) is 9.73 mm which is enough for most fastener evaluation. When larger fasteners need to be evaluated, more measurement spots or some spot spacing can be included with a change in the programming of the scanner control voltage matrix. For the present line scan of 512 spots, it takes 22 ms. An *X*-*Y* perpendicular two-line scan takes 44 or 90 ms if including OCT image processing. This speed is fast enough for handheld fastener scan measurement without the concern of hand instability provided that the handheld scanner is resting on the measurement surface with a soft plastic or rubber ring spacer. If we need to perform multi-line C-scan of 128 lines of 128 spots per line, it takes 2.2 seconds. A 256 × 256

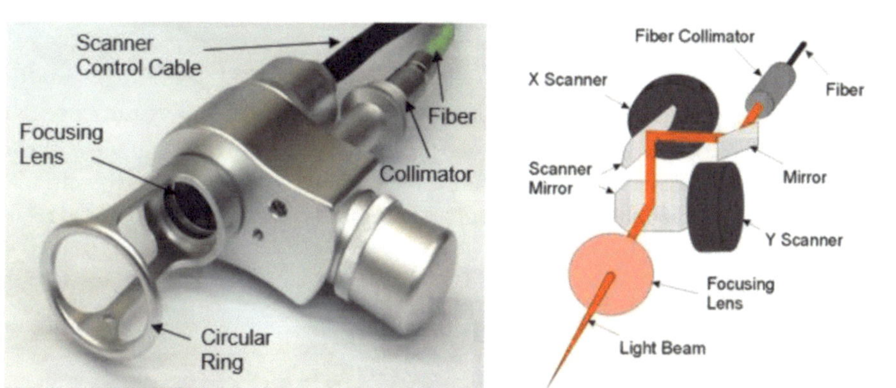

Figure 2.
Picture and schematic of the SD-OCT handheld scanner.

C-scan takes about 7.6 seconds. This speed, which includes the OCT image processing time, is also reasonably fast.

The axial measurement resolution is set by the spectral bandwidth of the SLD light source. Due to the non-Gaussian spectral distribution of the SLD light source, the present axial resolution is determined to be 8.5 μm in the air, which is better than most laser light scan measurements [2]. If using a 100 nm bandwidth light source, the SD-OCT axial resolution could be improved to 4.3 μm in the air. The axial resolution can be further improved to ~3 μm by using a 150 nm bandwidth light source instead. The lateral resolution is currently set by the focused beam spot size to be 19 μm, which is enough for fastener flushness evaluation. The lateral resolution may be improved by using shorter focal length lens on the scanner with the trade-off of a smaller FOV or by using a super-resolution technique with the trade-off of a longer scanning and image processing time [17].

3. Evaluation of fastener flushness

To perform fast fastener flushness evaluation using the SD-OCT without demanding center alignment of the scanner and fastener, we developed an X-Y perpendicular two-line scan pattern to determine the fastener's center position, radius, relative tilting angle, and flushness with the surrounding surface. As schematically illustrated in **Figure 3**, the red X-Y scan line is off-center. With the determination of fastener edge coordinate information $(x_1, 0, h_{x1})$, $(x_2, 0, h_{x2})$, $(0, y_1, h_{y1})$, and $(0, y_2, h_{y2})$, where height h is in respect to surrounding surface, we can determine the fastener surface center and height as.

$$\left(x_{center}, y_{center}, h_{center}\right) = \left(\frac{x_1 + x_2}{2}, \frac{y_1 + y_2}{2}, \frac{h_{x1} + h_{x2}}{2} + \frac{\left(h_{y2} - h_{y1}\right)\left(y_1 + y_2\right)}{2\left(y_2 - y_1\right)}\right) \quad (1)$$

The height of the fastener surface center position is calculated under the assumption that the fastener head is a flat circle. Hence, the height does not account for the cavity. Using the center position calculated above, the radius of the fastener can also be determined as

$$R = \sqrt{\left(\frac{x_2 - x_1}{2}\right)^2 + y_{center}^2 + \left(h_{center} - h_{x1}\right)^2} \quad (2)$$

By using the four edge coordinates established earlier, two vectors can be generated for the two scan lines. With these two vectors, taking the cross product will result in the normal vector, $\hat{N}$. Taking the dot product of this vector $\hat{N}$ with the normal vector of the surrounding surface $\hat{Z}$ will result in the tilting angle θ of the fastener. Since the scans conducted with the handheld scanner utilize the circular ring, it is assumed that the surrounding surface is flat. Thus, the tilting angle θ can be calculated as

$$\theta = cos^{-1}\left(\frac{\left(x_2 - x_1\right)\left(y_2 - y_1\right)}{\sqrt{\left(y_2 - y_1\right)^2\left(h_{x2} - h_{x1}\right)^2 + \left(x_2 - x_1\right)^2\left(h_{y2} - h_{y1}\right)^2 + \left(x_2 - x_1\right)^2\left(y_2 - y_1\right)^2}}\right) \quad (3)$$

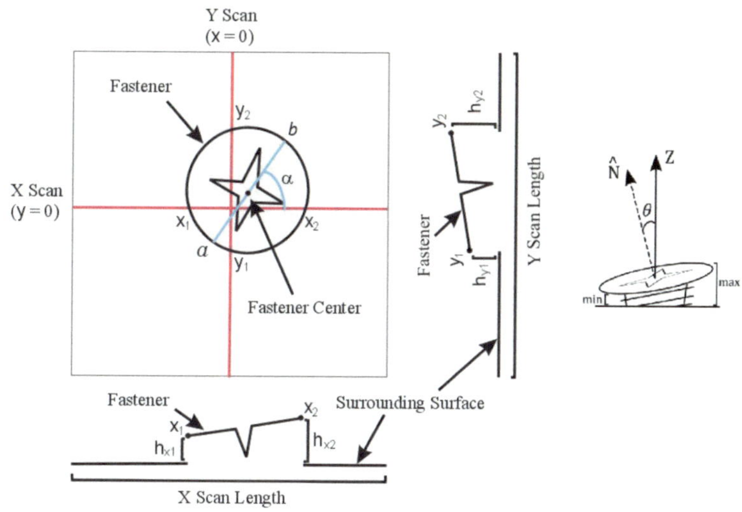

Figure 3.
Schematic for the measurement of fasteners.

To determine the flushness of the fastener, we can find the minimum and maximum heights h_a and h_b as shown in **Figure 3**. To do so, we first find the blue line slant angle α as

$$\alpha = tan^{-1}\left(\frac{(x_2 - x_1)(h_{y2} - h_{y1})}{(y_2 - y_1)(h_{x2} - h_{x1})}\right) \tag{4}$$

By using angle α, new edge coordinates (x_a, y_a) and (x_b, y_b) can be determined:

$$(x_a, y_a) = (x_{center} - Rcos(\theta)\cos(\alpha), y_{center} - Rcos(\theta)\sin(\alpha)) \tag{5}$$

$$(x_b, y_b) = (x_{center} + Rcos(\theta)\cos(\alpha), y_{center} + Rcos(\theta)\sin(\alpha)) \tag{6}$$

Knowing these two points a and b, the minimum and maximum heights of the fastener with the surrounding surface can be calculated. The equations to calculate the height for both points are

$$h_a = h_{x1} + \frac{h_{x2} - h_{x1}}{x_2 - x_1}x_a + \frac{h_{y2} - h_{y1}}{y_2 - y_1}y_a \text{ and } h_b = h_{x1} + \frac{h_{x2} - h_{x1}}{x_2 - x_1}x_b + \frac{h_{y2} - h_{y1}}{y_2 - y_1}y_b \tag{7}$$

To demonstrate the effectiveness of the fastener flushness evaluation, we acquired SD-OCT *X-Y* scan images of a fastener as shown in **Figure 4**. To minimize image line saturation and formation of image echo, the light source power and thus the image brightness are reduced. Based on the off-center information of the images, we obtained the coordinates x_1, x_2, y_1, and y_2 and flushnesses h_{x1}, h_{x2}, h_{y1}, and h_{y2}. Using the measurement technique mentioned above, we obtained the fastener radius of 3.53 mm which is close to the actual radius of 3.54 mm. The measured minimum and maximum flushness were 244.4 and 549.3 μm, respectively. The determined tilting angle θ and α were 2.47° and −19.45°, respectively. **Figure 5** shows a C-scan 3D image of the same fastener [18] acquired by 256 × 256 scan matrix showing the capability of the SD-OCT system. The saved and

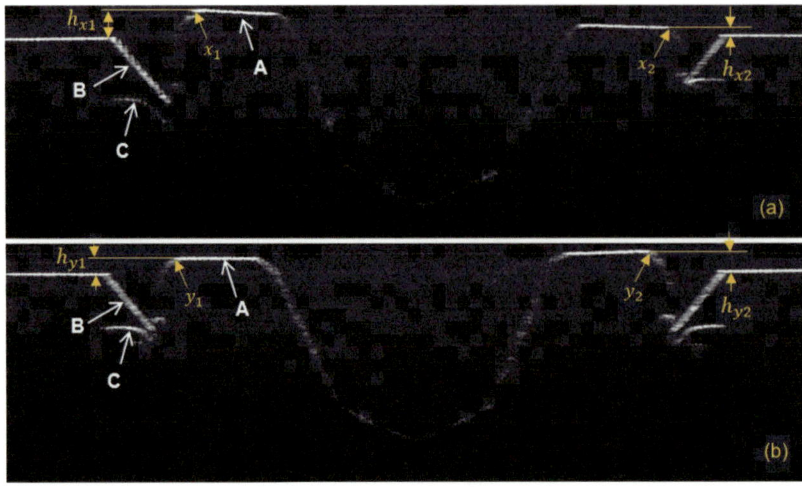

Figure 4.
OCT X-Y scan image of a fastener with scanner center not aligned to the fastener center. (a) Image of X scan and (b) image of Y scan. The lateral scan range is 12 mm and the scan depth is 3.65 mm in the air. A is the fastener, B is the countersink, and C is the reflection image of the fastener edge by the countersink.

Figure 5.
3D OCT image of the fastener [18].

documented 3D image data can be extracted to obtain any portion of the B-scan image for inspection reporting.

4. Evaluation of countersink surface profile

The seating of the fastener would depend on the quality of the countersink and underlying structure. For a cone shape countersink, it is important to evaluate its slant surface angle β as shown in **Figure 6**. To do so, we performed X-Y perpendicular two-line scan with scan center to be on any part of the cone shape slant surface without scanning through the countersink center screw hole. This can be ensured by not seeing a deep hole in the two line scan images. Instead, we should observe a curved scan surface line in each X and Y line scan image as illustrated in **Figure 6**. By taking the derivative on the X and Y scan data at the scan center O, we obtain

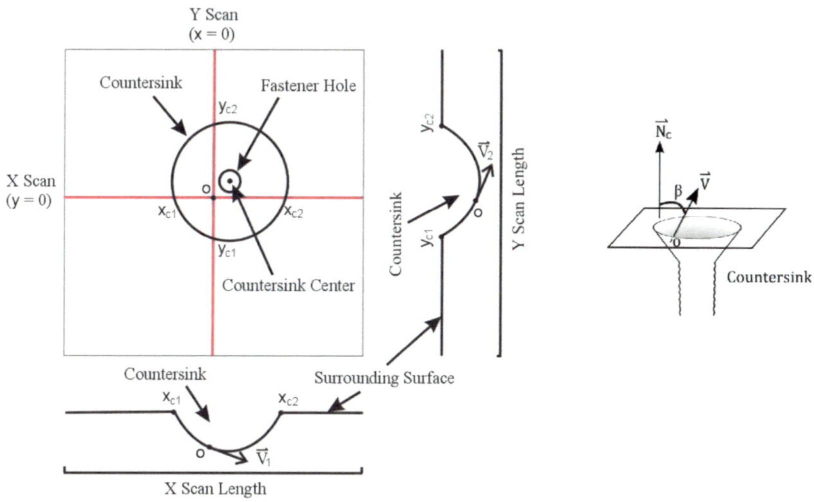

Figure 6.
Schematic of X-Y *line scan measurement of a countersink.*

two slopes and thus two tangential vectors $\vec{V}_1$ and $\vec{V}_2$. The cross product of the two vectors gives us the countersink slant surface normal vector $\vec{V}$.

This surface normal vector $\vec{V}$ is obtained from the X and Y scan data which is in reference to the handheld scanner's circular ring surface. This surface may be not exactly the same as the countersink outside surface. To determine the outside flat surface normal $\vec{N}_c$ for slant surface angle β determination, we first determine the countersink edge coordinates $(x_{c1}, 0, z_{xc1})$, $(x_{c2}, 0, z_{xc2})$, $(0, y_{c1}, z_{yc1})$, and $(0, y_{c2}, z_{yc2})$. Then the surface normal vector $\vec{N}_c$ is obtained.

$$\vec{N}_c = \langle (y_{c2} - y_{c1})(z_{xc1} - z_{xc2}), (x_{c2} - x_{c1})(z_{yc1} - z_{yc2}), (x_{c2} - x_{c1})(y_{c2} - y_{c1}) \rangle \quad (8)$$

Using this vector and the previously found surface normal vector, the slant angle of the countersink surface can be calculated by

$$\beta = cos^{-1}\left(\frac{\vec{V} \cdot \vec{N}_c}{|\vec{V}||\vec{N}_c|}\right) \quad (9)$$

Similar to the equation used to calculate the radius of the fastener head, the countersink radius can be calculated as

$$R_c = \sqrt{\left(\frac{x_{c2} - x_{c1}}{2}\right)^2 + \left(\frac{y_{c2} + y_{c1}}{2}\right)^2 + \left(\frac{z_{xc1} + z_{xc2}}{2} + \frac{(z_{yc2} - z_{yc1})(y_{c1} + y_{c2})}{2(y_{c2} - y_{c1})} - z_{xc1}\right)^2} \quad (10)$$

To demonstrate the effectiveness of countersink surface profile evaluation, we acquired SD-OCT X-Y scan images of a countersink as shown in **Figure** 7. The countersink slant surface angle β and radius R_c were found to be 50.9° and 4.59 mm

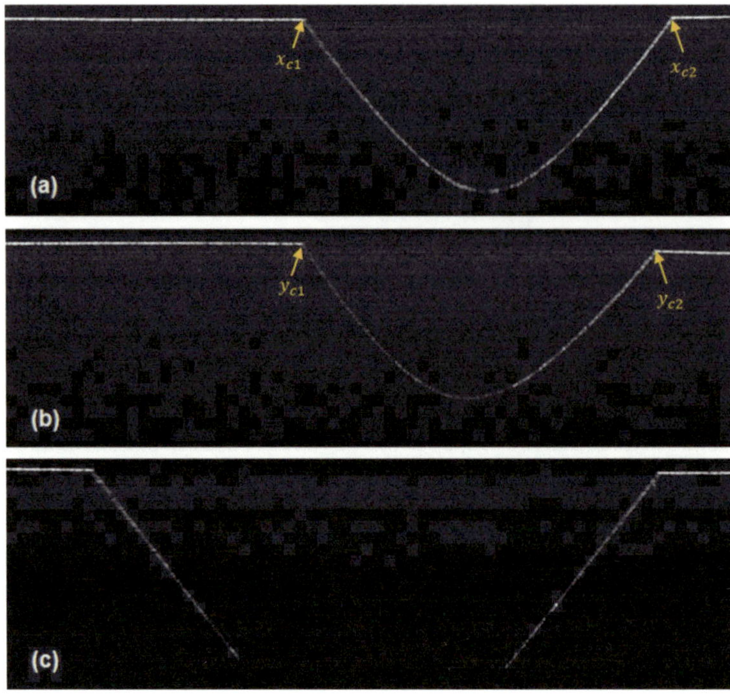

Figure 7.
OCT X-Y scan image of a countersink with scanner center on any part of a countersink slant surface. (a) Image of X scan and (b) image of Y scan with lateral scan range of 16 mm and scan depth of 3.65 mm in the air. The countersink edge coordinates are x_{c1}, x_{c2}, y_{c1}, and y_{c2}. (c) Through center scan image of the countersink with lateral scan range of 12 mm.

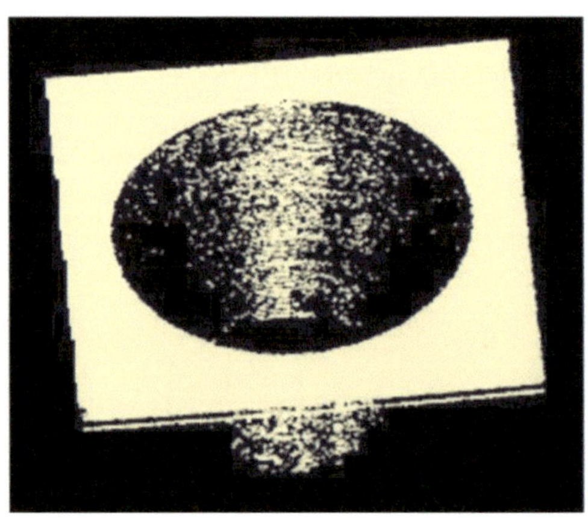

Figure 8.
3D OCT image of the countersink [18].

close to the actual value of 50.5° and 4.57 mm, respectively. **Figure 8** shows a C-scan 3D image of the same countersink [18] acquired by a 256 × 256 scan matrix.

5. Conclusion

Considering that high–speed vehicles and aircraft demand high-quality inspection of fastener flushness and countersinks, the SD-OCT scanner would be a valuable inspection device. Using an *X-Y* scanning pattern with vector calculation not only avoids center alignment of the scanner and fastener but also allows for quick evaluation of the fastener's radius, tilt angle, and flushness. Similarly, countersink evaluation also uses the X-Y scanning pattern to determine the countersink's radius and surface angle. In addition, the SD-OCT device is capable of producing 3D surface profiles by using multi-line C-scans. With all of these features, its high axial resolution on top of all these measurements show that this device has potential for practical use in fastener flushness and countersink evaluation.

Conflict of interest

The authors declare no conflict of interest.

Author details

James H. Wang[1] and Michael R. Wang[2]*

1 New Span Opto-Technology Inc., Miami, FL, USA

2 Department of Electrical and Computer Engineering, University of Miami, Coral Gables, FL, USA

*Address all correspondence to: mwang@miami.edu

References

[1] Alberts DG. Countersink Depth Gauge. Patent US5758433; 1998

[2] Zhang H, Tao W, Liu M, Zhao H. A laser scanning system for the inspection of fasteners in railways. Lasers in Engineering. 2013;**24**:229

[3] Ren SW, Liu WW, Liu H, Liu MH, Liu WH, Yang JF, et al. Tao, Structured light-based high-speed detection system and method for railway fastener, patent CN102221553A; 2011

[4] Huang D, Swanson EA, Lin CP, Schuman JS, Stinson WG, Chang W, et al. Optical coherence tomography. Science. 1991;**254**:1178-1181. DOI: 10.1126/science.1957169

[5] Izatt JA, Hee MR, Swanson EA, Lin CP, Huang D, Schuman JS, et al. Micrometer-scale resolution imaging of the anterior eye in vivo with optical coherence tomography. Archives of Ophthalmology. 1994;**112**:1584. DOI: 10.1001/archopht.1994.0109024009 0031

[6] Amaral MM, Raele MP, Caly JP, Samad RE, Vieira ND, Freitas AZ. Roughness measurement methodology according to DIN 4768 using optical coherence tomography (OCT). Proceedings of SPIE. 2009;**7390**: 73900Z. DOI: 10.1117/12.827748

[7] Darlow LN, Akhoury SS, Connan J. A review of state-of-the-art speckle reduction techniques for optical coherence tomography fingertip scans. Proceedings of SPIE. 2015;**9445**:944523. DOI: 10.1117/12.2180537

[8] Akhoury SS, Darlow LN. Extracting subsurface fingerprints using optical coherence tomography. IEEE Xplore. 2015:184

[9] Li Z, Meng Z, Wang L, Liu T, Yao SX. Tomographic inspection of fiber coils using optical coherence tomography. IEEE Photonics Technology Letters. 2015;**27**:549. DOI: 10.1109/LPT.2014.2384837

[10] Cerrotta S, Morel EN, Torga JR. Scanning optical coherence tomography applied to the characterization of surfaces and coatings. Procedia Materials Science. 2015;**9**:142. DOI: 10.1016/j.mspro.2015.04.018

[11] Hutiu G, Duma VF, Demian D, Bradu A, Podoleanu AG. Surface imaging of metallic fractures using optical coherence tomography. Applied Optics. 2014;**53**:5912. DOI: 10.1364/AO.53.005912

[12] Meemon P, Yao J, Lee K-S, Thompson KP, Ponting M, Baer E, et al. Optical coherence tomography enabling nondestructive metrology of layered polymeric GRIN material. Scientific Reports. 2013;**3**:1709. DOI: 10.1038/srep01709

[13] Kim JS, Ishikawa H, Gabriele ML, Wollstein G, Bilonick RA, Kagemann L, et al. Retinal nerve fiber layer thickness measurement comparability between time domain optical coherence tomography (OCT) and spectral domain OCT. IVOS. 2010;**51**:896. DOI: 10.1167/iovs.09-4110

[14] de Boer JF, Cense F, Hyle Park B, Pierce MC, Tearney GJ, Bouma BE. Improved signal-to-noise ratio in spectral-domain compared with time-domain optical coherence tomography. Optics Letters. 2003;**28**:2067. DOI: 10.1364/OL.28.002067

[15] Wojtkowski M. High-speed optical coherence tomography: Basics and applications. Applied Optics. 2010;**49**: D30. DOI: 10.1364/AO.49.000D30

[16] Drexler W, Liu M, Kumar A, Kamali T, Unterhuber A, Leitgeb RA.

Optical coherence tomography today: Speed, contrast, and multimodality. Journal of Biomedical Optics. 2014;**19**: 071412. DOI: 10.1117/1.JBO.19.7.071412

[17] Shen K, Liu H, Baig S, Wang MR. Improving lateral resolution and image quality of optical coherence tomography by the multi-frame superresolution technique for 3D tissue imaging. Biomedical Optics Express. 2017;**8**:4887. DOI: 10.1364/BOE.8.004887

[18] Wang JH, Wang MR. Handheld non-contact evaluation of fastener flushness and countersink surface profiles using optical coherence tomography. Optics Communications. 2016;**371**:206. DOI: 10.1016/j.optcom.2016.03.069